BLUE PLANET

A PORTRAIT OF EARTH

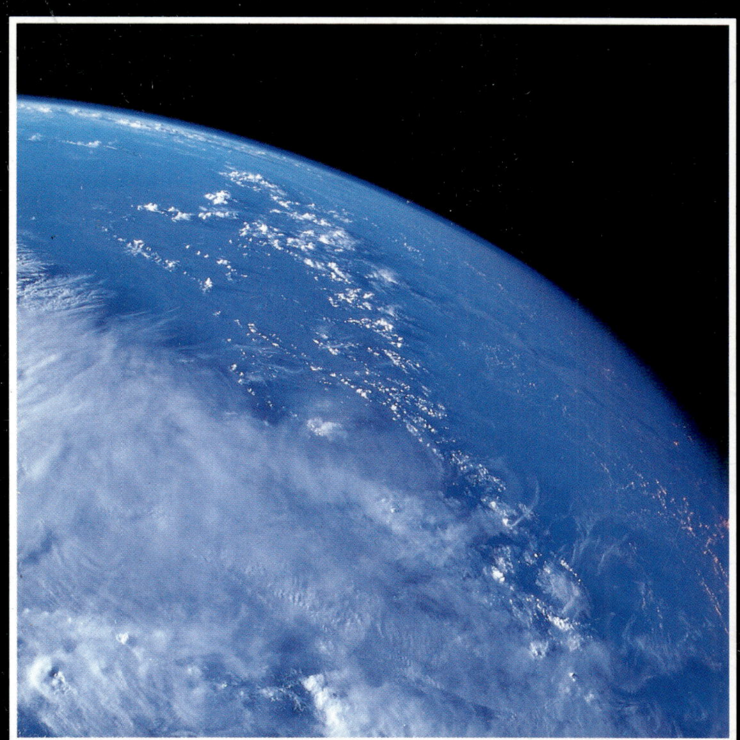

BY LYDIA DOTTO

Foreword by Martin Harwit
Director, National Air and Space Museum

A Sue Katz and Associates, Inc., Book, in association with the
Smithsonian Institution, Lockheed Corporation, Imax Corporation, and Harry N. Abrams, Inc.

Editor: Sharon AvRutick
Designer: Dana Sloan
Imax Project Manager: Susan Mander
Imax Photo Editor: Gayle Bonish

Library of Congress Cataloging-in-Publication Data
Dotto, Lydia.
   Blue planet: a portrait of earth /
by Lydia Dotto; foreword by Martin Harwit.
      p.   cm.
   ISBN 0-8109-2472-2
   1. Pollution—Pictorial works.   2. Earth—Photographs from space.
I. Title.
TD174.D68   1991
363.73′2′0222—dc20   90–48072
                                                CIP

Copyright © 1991 Smithsonian Institution/Lockheed Corporation
Published in 1991 by Harry N. Abrams, Incorporated, New York
A Times Mirror Company
All rights reserved. No part of the contents of this book may be reproduced
without the written permission of the publisher

IMAX® is a registered trademark of Imax Corporation, Toronto, Canada.
Blue Planet® is a registered trademark of the Smithsonian Institution and
Lockheed Corporation.

Printed and bound in Japan

PHOTO CREDITS

All images in this book are © 1990 Smithsonian Institution/Lockheed Corporation, except for page 33 top: Landsat™, Courtesy of Pat Chavez, U.S.
Geological Survey; page 38: NASA; page 44: Volcano Film Partner-
ship/Courtesy of Graphic Films Corporation, Reuben H. Fleet Space Theater
and Science Center, Science Museum of Minnesota, Museum Education
Productions Inc.; page 58–59: © 1985 W. T. Sullivan; and pages 6 right, 7, 13,
and 17 bottom: NASA.

Imax would also like to thank the following for their role in obtaining images:
Gerald Calderon, IFREMER (French Research Institute for Ocean Explora-
tion); Digital Image Animation Laboratory, NASA's Jet Propulsion Laboratory,
Pasadena, California; and Media Services Corporation, NASA, Lyndon B.
Johnson Space Center, Houston, Texas.

These images were filmed in IMAX by the astronauts of Shuttle Missions 29,
31, 32, 34, and 61B. Ground scenes were filmed by Ben Burtt, David Douglas,
Graeme Ferguson, Ivan Galin, Greg MacGillivray, James Neihouse, Brad
Ohlund, Neil Rettig, Gail Singer, and Adrian Warren. The film *Blue Planet* was
produced by Graeme Ferguson, with Toni Myers as Writer–Editor–Narrator
and Phyllis Ferguson as Associate Producer.

METRIC EQUIVALENTS

1 centimeter = .39 inches

1 meter = 100 centimeters = 3.28 feet

1 kilometer = 1000 meters = .62 miles

1 square kilometer = .39 square miles = 247 acres

When Buzz Aldrin, Neil Armstrong, and Mike Collins returned home to earth after humanity's first successful landing on the moon, the word *home* took on a broadened meaning. Home is where we are born and live our formative years. Home is where we wish to return after long absences, to make sure it's still there. Home now also became our planet earth. Only by leaving to visit an inhospitable moon could we fully realize how much our own planet meant to us.

We recognize the need for change and often welcome great leaps forward. But some things should be kept apart, saved just as they are, just as they always have been. Our home planet, the blue planet, the third of a set of nine otherwise-lifeless planets orbiting the sun, is one of these.

Not that earth has been changeless over the aeons. Four billion years ago, a visitor here would have found a desolate world endlessly orbiting the sun. Then, simple life-forms evolved, altering earth's surface and atmosphere, preparing the planet for an explosive proliferation of plant and animal life, with species numbering in the millions.

Throughout this time, enormous landmasses slid across earth's surface. Where they collided, their edges crushed to form chains of mountains. Volcanoes, pipelines from earth's semimolten interior, exploded their caps, pouring out molten lava. Mile-size asteroidal bodies pounded the earth at irregular intervals, spewing out soil and rock. Ice ages came and went every few thousand years, uprooting life-forms, forcing vast migrations north and south. Meanwhile, torrential rains swept in by hurricanes gathering above the oceans eroded the mountains and leveled the continents, only to have new mountain chains springing up elsewhere in response to continental drift.

Life on earth has survived those calamitous changes through its ability to adapt. For every change in our environment, changing life-forms arose, grasping advantage where others saw only destruction. Among these changes was the growth of intelligence.

Intelligence has permitted the harvesting of forests, the planting of crops, the domestication of animals, and a wide-ranging contract between humans and their planet. The basis of this pact at all times has been moderation. Life-forms would take their share of the wealth, and inanimate nature would continue on its destructive course — but neither would prevail.

Now, this contract is being threatened, with humanity emerging both as victor and victim. Our numbers and our industry have increased to a point where we are massively altering the environment, perhaps forever. Unwittingly, we are launched on an uncontrolled experiment that is changing our home, not knowing where our actions will take us or our planet.

From space, a first glance still shows a lovely blue sphere, a planet covered largely by oceans, protected by a thin shield of precious air. As we approach, we begin to see growing ozone holes in the atmosphere's polar regions and steadily increasing atmospheric carbon dioxide. Closer still, we discern vast smoke plumes rising from ravaged rain forests, land being prepared for farming. But those farms may never flourish; downslope from them, we see the soil, once held together by roots of trees, swept downstream in the rush of rivers plunging into the oceans.

Three years ago, the National Air and Space Museum began discussions with Producer Graeme Ferguson at Imax Corporation for a film that would show both the natural and the human forces as they act on our environment. Through the exceptionally clear footage obtained in large-format 70mm pictures from space, we felt we could make vivid the scale of the challenges we face in keeping our home planet hospitable for life as it has existed since earliest geological times. This seemed to us to be a very important and timely message to convey, but we knew that a successful film would require the aid of a large number of organizations and individuals. To our delight, we found enthusiastic help from all whom we approached.

The National Aeronautics and Space Administration made this film possible by providing access for our cameras aboard space-shuttle flights and by encouraging astronauts to film all of the footage obtained from space. The Lockheed Corporation, with whom the Museum and Imax had produced a previous film, became a partner once again, providing major financial support for this venture. And, to make certain that the film accurately reflected today's best scientific understanding of the human and natural forces at work, we sought advice from a panel of experts with wide-ranging insights. It included Dr. Ghassan Asrar of the National Aeronautics and Space Administration; Dr. D. James Baker, director of Joint Oceanographic Institutions, Inc.; Prof. Francis Bretherton of the University of Wisconsin; Brian Duff, Dr. Patricia Jacobberger, Dr. Ted Maxwell from the Museum; Toni Myers, the film's writer, editor, and narrator; and Barbara Valentino from the National Academy of Sciences. Dr. Steven Soter, from the Museum, who was part of the team, was also the main scientific advisor for this book. Our deepest thanks go to all of them.

Martin Harwit
Director, National Air and Space Museum
Washington, D.C.

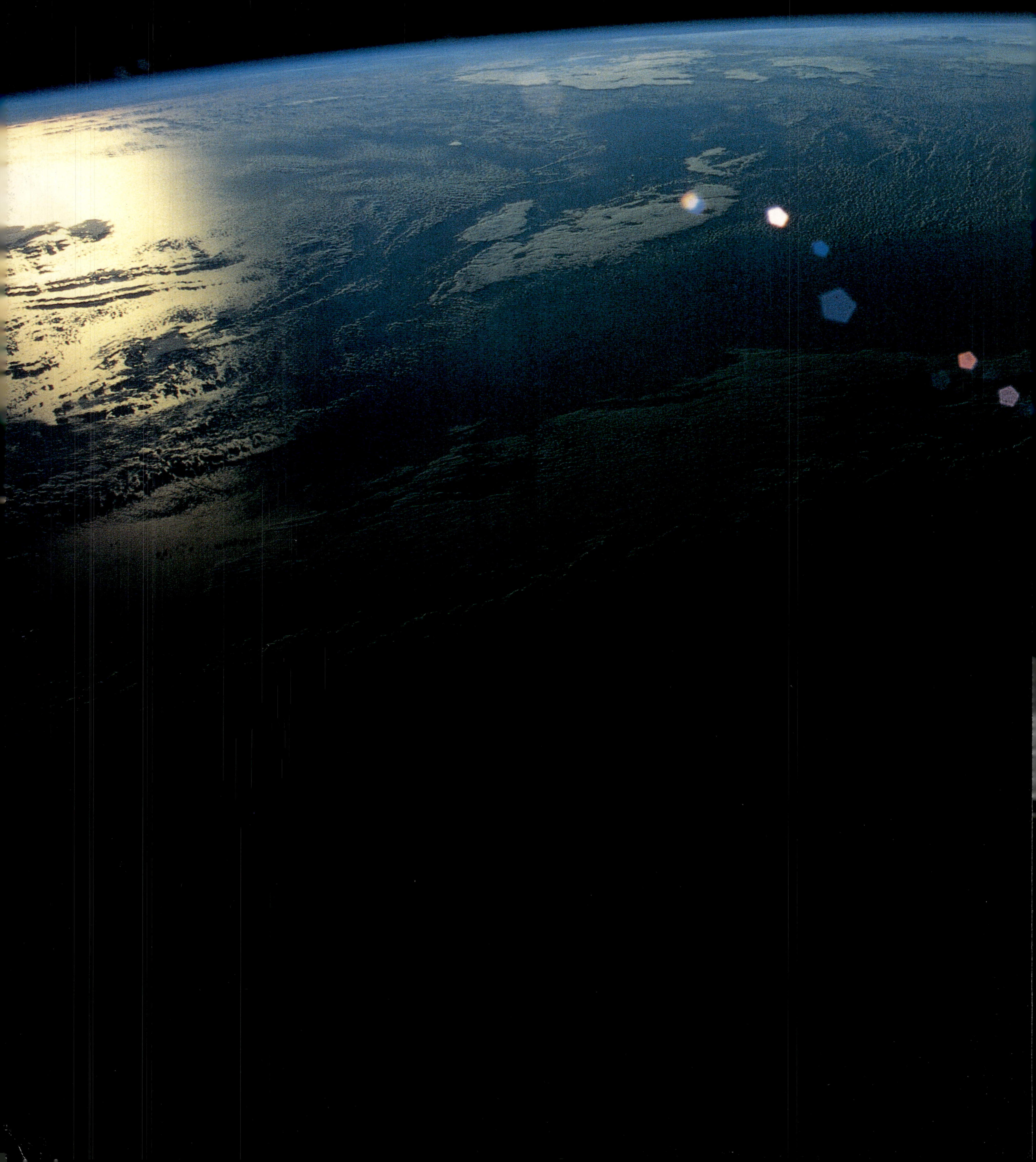

The line where the dayside of the earth
fades into the nightside is known as the
terminator. A space shuttle orbits the earth
about every ninety minutes and crosses the
terminator twice during each orbit. Thus,
astronauts on board the space shuttle
experience sixteen sunrises and sixteen
sunsets every twenty-four hours.

The technology that carries us away from earth provides us with a unique vantage point from which to look back at it. This has profoundly changed our perception of our planet, forcing us to confront new realities: first, that earth is indeed a planet—and a rather small one at that. Second, that it is very much alone, adrift in a black and empty space with no friendly neighbors nearby. And finally, that it is all we have. We know from what little spacefaring we've done so far that earth is unique in our solar system in its ability to nurture and sustain a vast array of living species. Since we've yet to find even a hint of life elsewhere, we must regard earth, for the foreseeable future, as a small haven in a mostly cold and lifeless universe.

Many young people today have never known a time when we did not have the unique view of earth that we get from space. Yet it's been barely more than two decades since the crew of Apollo 8, the first humans to venture beyond earth's orbit, turned their cameras homeward and provided us with those first electrifying pictures of a brilliant blue-and-white planet rising above the barren gray-brown surface of the moon.

Through the eyes and cameras of the astronauts who followed Apollo 8 to the moon, we witnessed time and again the stark contrast between the deadness of the lunar surface and the aliveness of the earth. Apollo 14 astronaut Edgar Mitchell, for example, described earth as "a sparkling blue-and-white jewel . . . like a small pearl in a thick sea of black mystery."

We saw—and perhaps truly appreciated for the first time—just what earth has provided us for millions of years: not just a home, but a life-support system. And we gained a deeper understanding of how much we have always taken for granted something that is remarkable, precious, and unique. Francis Bretherton, director of the Space Science and Engineering Center at the University of Wisconsin and a member of the *Blue Planet* scientific advisory committee, points out that human communication depends significantly on vision, so being able actually to *see* the earth as a whole had an important emotional impact. Photographs from space provide a "vicarious spaceflight" that shows us "things we half-knew but had never really realized before," he says.

It's hardly surprising then that images of earth from space have been widely credited with planting the seeds of our current environmental awareness and our growing concern for the fate of the planet. Indeed, our alarm about the future of our own species stems in part from lessons we've learned from going into space ourselves. Surviving out there is one of the most challenging things we've ever attempted, and it has given us a keen appreciation of just how difficult it is to keep humans—indeed, any-thing—alive once earth's protective cocoon has been left behind. Space provides none of the raw materials needed for survival: there is no food out there, no water, no air.

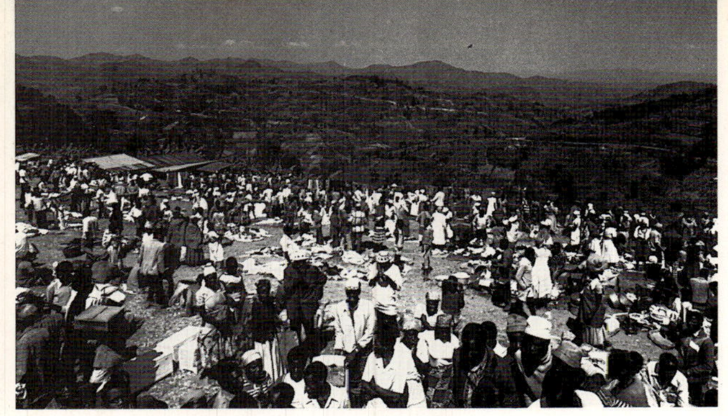

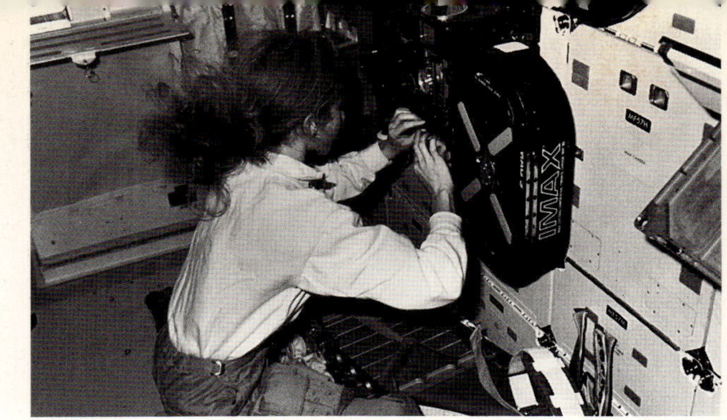

Consequently, we venture into space enclosed in a technological life-support system—a spacecraft—that can provide, in a limited way and for short periods of time, what the earth has provided for all life for hundreds of millions of years. In this way, a spacecraft is a microcosm of earth: a source of food, air, water, atmospheric pressure, and protection from the extremes of heat and cold found in space. Since these resources are severely limited in a spacecraft, we've learned to conserve and recycle them. We recirculate air and remove excess carbon dioxide from it so that the atmosphere can continue to support life. We carefully collect wastes and prevent dangerous chemicals from being released, because these would contaminate the life-support system and endanger the crew.

We're extraordinarily careful about taking these precautions in space—and yet we appear to be neglecting them on earth. Why? Perhaps because we haven't really thought of earth as a closed life-support system—or perhaps because we've always considered it too huge and resilient for us to do much damage. But with the new perspective that we've obtained from space, we can see more clearly that earth is actually quite small, that its resources are finite, that it has boundaries and limits—and that we are capable of pushing it to those limits and perhaps beyond.

This is a hard lesson to learn, and it's not surprising that we still have trouble accepting it. After all, for most of human history, we have regarded this planet—with its great landmasses and even greater oceans, its plentiful forests and rivers, its great diversity of plants and animals, and its protective atmosphere—as inexhaustible and unassailable. And, indeed, when our species was small in numbers, technologically unsophisticated, and limited in its range of activities, this perception was probably true. For most of the approximately two million years we have been on earth, we lived in successful harmony with nature.

HUMAN BEINGS AS A PLANETARY FORCE

It is only within the last 150 years, since the Industrial Revolution began, that our numbers and our use of technology began pushing the global life-support system too hard and too fast. In the last half of the twentieth century, we have become what environmentalists call a "planetary force."

In 1800, there were only about one billion people on earth; it took until 1930 for that number to double, but only until 1975 for it to double again. By the beginning of 1990, the human population stood at 5.2 billion and the growth rate had reached nearly one billion per decade—the fastest in human history.

Human beings consume large amounts of energy and resources and produce tremendous amounts of waste—much of it in the form of complex, deadly chemicals that have never existed in nature and that can persist in the environment for decades or centuries or even longer. Humans alter the face of the earth with farms and cities, roads and airports; we use rivers and oceans as dumps for garbage and sewage; we emit gases that change the fundamental chemistry of the atmosphere, perhaps irreversibly. We are, in other words, engaged in an exercise on earth that would surely be fatal if we were to do it in space.

There are many uncertainties about what will happen to the global environment over the next century, and how that will affect and be affected by human activities, but many environmentalists are now convinced that we are risking our own viability, and that of many other living things, by seriously undermining the biological carrying capacity of planet earth. In 1988, scientists from all over the world attending a climate conference in Toronto stated that earth's atmosphere is being changed at an unprecedented rate by human activities and population growth and our wasteful use of energy. In a statement remarkable for its forcefulness, the scientists warned that "humanity is conducting an unintended, uncontrolled, globally pervasive experiment whose ultimate consequences could be second only to nuclear war."

What we see of the earth from space lends weight to these concerns. On the night side of the planet, whole continents are etched out in light, their great cities blazing like beacons—a graphic testament not only to the patterns of human energy consumption, but also to the sheer extent of it. Like a giant Rorschach test, these gilt-edged maps can be interpreted in different ways: while some may see them as a symbol of progress and improvement in the human condition, others see them as a disquieting reminder of the way in which human beings have encroached upon, and indeed overwhelmed, earth's natural ecosystems.

In fact, human activity has left its mark on the earth in many ways that are clearly visible from space—a fact that took astronauts somewhat by surprise in the early days of the space program. Astronaut James F. Buchli, a veteran of three shuttle flights who has logged 243 hours in space, comments that "the eye really integrates things in a way you'd almost think isn't possible." For example, he says that the contrails of aircraft and the wakes of ships can sometimes be seen from orbit. But it is natural phenomena, like sunrises and sunsets, that provide the most memorable images. "Watching a sunrise unfold from orbit is the most spectacular event. You go from entirely black to brilliant in a matter of minutes, and every shade in the spectrum comes in for a short time."

Earth watching is an astronaut's favorite pastime. Buchli was awed on one mission to see the Grand Canyon covering an entire shuttle window: "It was really something—you could just see the amount of water

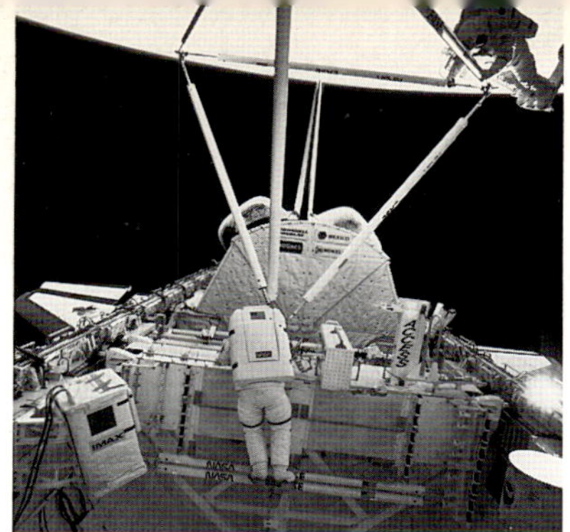

and energy and erosion needed to make that happen." On another occasion, he witnessed the aurora australis, the Southern Hemisphere's equivalent of the northern lights. It appeared, Buchli remembers, as a "huge octopus or starfish of light over the South Pole, with its tentacles coming out." He was impressed with how high the curtains of light extended; they were clearly visible through the shuttle windows and even though the vehicle was high above them, "it seemed like the orbiter was immersed in these translucent sheens of light."

More than the astronauts' words, however, we have the pictures they've taken over nearly three decades of spaceflight. Many astronauts have been avid photographers and more than three dozen have been trained to use the large-format IMAX camera, which has been flown on eight shuttle missions so far. The crew of five missions—STS 61-B in 1985, STS-29 and STS-34 in 1989, and STS-31 and STS-32 in 1990—filmed the footage seen in *Blue Planet*. The unique IMAX technology has enabled them to capture on film at least a part of the spaceflight experience to share with those of us who will never have the opportunity to see the earth from space. Many astronauts have commented that IMAX films—shown on screens up to eight stories tall and 33.5 meters wide—provide millions of people with an experience as close to being in orbit with them as one can possibly get.

The space program has also given us other tools for observing our planet—satellites that can see and measure things the human eye cannot. Satellite instruments can detect wavelengths beyond the visible range, they can "see" through clouds and at night, and they can monitor concentrations of atmospheric chemicals and the ways in which they are changing. Moreover, satellites provide more frequent and more comprehensive coverage of the earth's surface than we can achieve any other way. Francis Bretherton points out that space-based observing is the only way to "cover the world quickly in a relatively uniform manner. We can get comparable data from places widely separated on the earth's surface." In fact, the 1987 report of the World Commission on Environment and Development, known as *Our Common Future*, noted that "outer space can play a vital role in ensuring the continued habitability of the earth, largely through space technology, to monitor the vital signs of the planet and aid humans in protecting its health."

Bretherton says that humans and satellites make "distinct contributions that are complementary to each other." He notes that most of our information comes from unmanned spacecraft, which cover the earth more often and more completely than manned vehicles, but human observers have been able to find things that instruments have not—interesting structures in the oceans and clouds, for example. "A trained human being sees patterns that are very hard to train a machine to see," he points out. "An observer in the shuttle can direct attention to something that might otherwise get missed."

Over the past two decades, observing the earth from space has resulted in an explosive increase in our ability to monitor and understand what is happening to the planet. It has also produced a few mysteries and surprises and much disquieting evidence about the state of the global environment.

Space images have much to tell us about the history of the earth. There's ample evidence that the planet has had a hard and at times violent past: its face is weathered and wrinkled, scarred with craters caused by impacting meteorites, with mountain chains uplifted by the collision of huge moving plates, with rifts torn open by earthquakes, and with canyons laboriously carved out of the rocks by rivers over millennia. Even after 4.5 billion years, earth remains a dynamic place and its face continues to be rearranged by such geological forces as earthquakes and volcanoes.

From space, we can see ample evidence of the impact of volcanic activity, past and present. The Hawaiian Islands, for example, were formed by the outflowing of lava from the ocean floor. Hawaiian-type volcanoes are the world's largest volcanic formations. Two on the Big Island—Mauna Loa and Mauna Kea—are actually the tallest mountains on earth; measured from the seafloor, they exceed the height of Mount Everest, which is 8,848 meters. Mauna Kea is the tallest, at about 9,750 meters above the ocean floor. It is extinct, and telescopes have been placed at its summit, which is 4,170 meters above sea level, taking advantage of the altitude to escape the obscuring effects of the earth's atmosphere. Things are somewhat less peaceful over on Mauna Loa and nearby Kilauea; they are among the largest active volcanoes on earth and regularly send fiery rivers of lava flowing down the mountainside, destroying roads, trees, and houses in their path. The nearby Kau Desert contains many strange lava formations.

Volcanoes have many personalities. Those in Hawaii overflow with lava that burns its way relentlessly to the sea. Other types of volcanoes literally blow their top, and these are not hard to see from space; they have often been caught in the act by an astronaut's camera. "The smoke or ash plume is very distinct," says James Buchli. On shuttle mission STS-34, in October, 1989, the crew captured a plume of ash trailing out of Sakurajima, a volcano in southern Japan, on film for *Blue Planet*.

Volcanoes cause a great deal of devastation in their immediate vicinity, but they can also have much more far-reaching effects. Violent

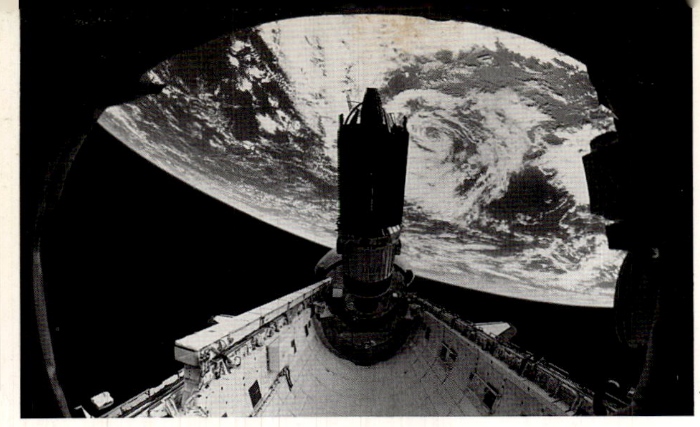

eruptions can inject huge amounts of ash, dust, and gases into the upper atmosphere and, as they spread around the world, they can block the sun's energy and thus affect the climate. In April, 1815, the Indonesian volcano Tambora exploded, ejecting so much dust and ash into the atmosphere that an area 450 kilometers to the west was shrouded in near-darkness for three days. (By way of comparison, the Tambora eruption ejected about one to two hundred times as much material as the 1980 Mount St. Helens eruption in Washington State.) Some scientists have speculated that Tambora may have been responsible for the following year being one of the coldest in several centuries. (In New England, which experienced untimely summer frosts, it was called "the year without a summer.") The eruptions of the Indonesian volcanoes Krakatoa in August, 1883, and Agung in 1963 were also followed by large-scale cooling; scientists believe this resulted, at least in part, from the massive injection of volcanic particles and gases into the upper atmosphere.

Studying the movement and behavior of volcanic ejecta has helped scientists trying to project the potential climatic consequences of a nuclear war. Their computer calculations suggest that, if a full-scale war were to occur, dust and soot blasted into the atmosphere by nuclear explosions and resulting fires would likely cause significant climatic disruptions on a global scale, possibly including a severe nuclear winter in some regions.

Although scientists remain concerned about the climatic consequences of nuclear war, in recent years they have focused their attention more urgently on what everyday human activities are doing to the earth's climate and weather. Here again, having an observing post in space has provided significant benefits; for example, it enables scientists to see large weather systems in their entirety, to track their movements, and to observe how they change over time. From space, we are entranced by the sight of billowy white clouds blanketing the earth in lacy curtains or winding themselves into intricate whirlpool patterns. These images have an almost ethereal quality, but their serenity is deceptive, for these clouds usually cause unsettled—and often very severe—weather at ground level.

For example, huge spiral cloud patterns, which can be large enough to blanket most of a continent, provide the "basic engine" that creates most midlatitude weather, according to University of Toronto physicist and climate expert Richard Peltier. These "cyclonic" structures are formed by warm and cold air wrapping around each other "like spaghetti on the end of a fork." Along the spiral's arms are warm and cold fronts, where most of the weather that affects our daily lives takes place, particularly unsettled and violent weather such as blizzards, thunderstorms, hurricanes, and tornadoes. One of the most impressive sights from space is that of a line of towering thunderheads stretching for hundreds of miles along a front.

Satellites have greatly improved our ability to predict hurricanes. "Hurricanes form over the wide spaces of the ocean and move from there across the coastline. It might take three, four, or five days before they reach land," Bretherton says. Spacecraft provide a unique platform for us "to watch these things from space, to know where they are and how they are forming, so we can say with much more confidence that a hurricane is coming and this is where it's probably coming ashore."

Space observations are also helping scientists gain a better understanding of clouds and the role they play, not only in creating short-term weather but in regulating longer-term climate. Bretherton says that pictures taken by shuttle astronauts have revealed interesting and puzzling structures in clouds that are "a real challenge to explain."

Clouds have contradictory effects on the heat balance of the earth: they warm the atmosphere by trapping heat beneath them, but they also have a cooling effect by reflecting the sun's energy back into space. Satellite data have shown there is tremendous variety in clouds; for example, clouds over the tropics are different from those over temperate zones, while those over oceans are different from those over land. Our awareness of these differences can be directly attributed to global space-based observing; before the advent of satellites, scientists had primarily studied the most accessible clouds, those in midlatitudes over land.

Scientists are trying to incorporate their new knowledge about cloud structures and patterns into the computer models used to predict the weather and longer-term climatic changes. These models are an attempt to recreate the earth inside a computer; complex mathematical equations are used to describe a system—for example, the earth's atmosphere—and to define the relationships and interactions among the various elements that affect it. These equations are then "run forward" to explore how the system might change over time. This technique is used not only to generate weather forecasts, but to create "what if" scenarios—for example, to project what energy use and the release of industrial chemicals might do to change the earth's climate in the future.

These models depend critically on the amount and accuracy of the information about the real world that's fed into them, but it's difficult and costly to get accurate information about what's going on everywhere; the oceans, for example, are still relatively poorly monitored. Observing the earth from space has helped considerably in improving these models and our understanding of the atmosphere. Although data from ground-based stations are still needed, it's impossible to beat space-based observations for the sheer amount and frequency of coverage and, in recent years, improvements in technology have permitted scientists to monitor climatic and environmental changes on the earth's surface in remarkable detail.

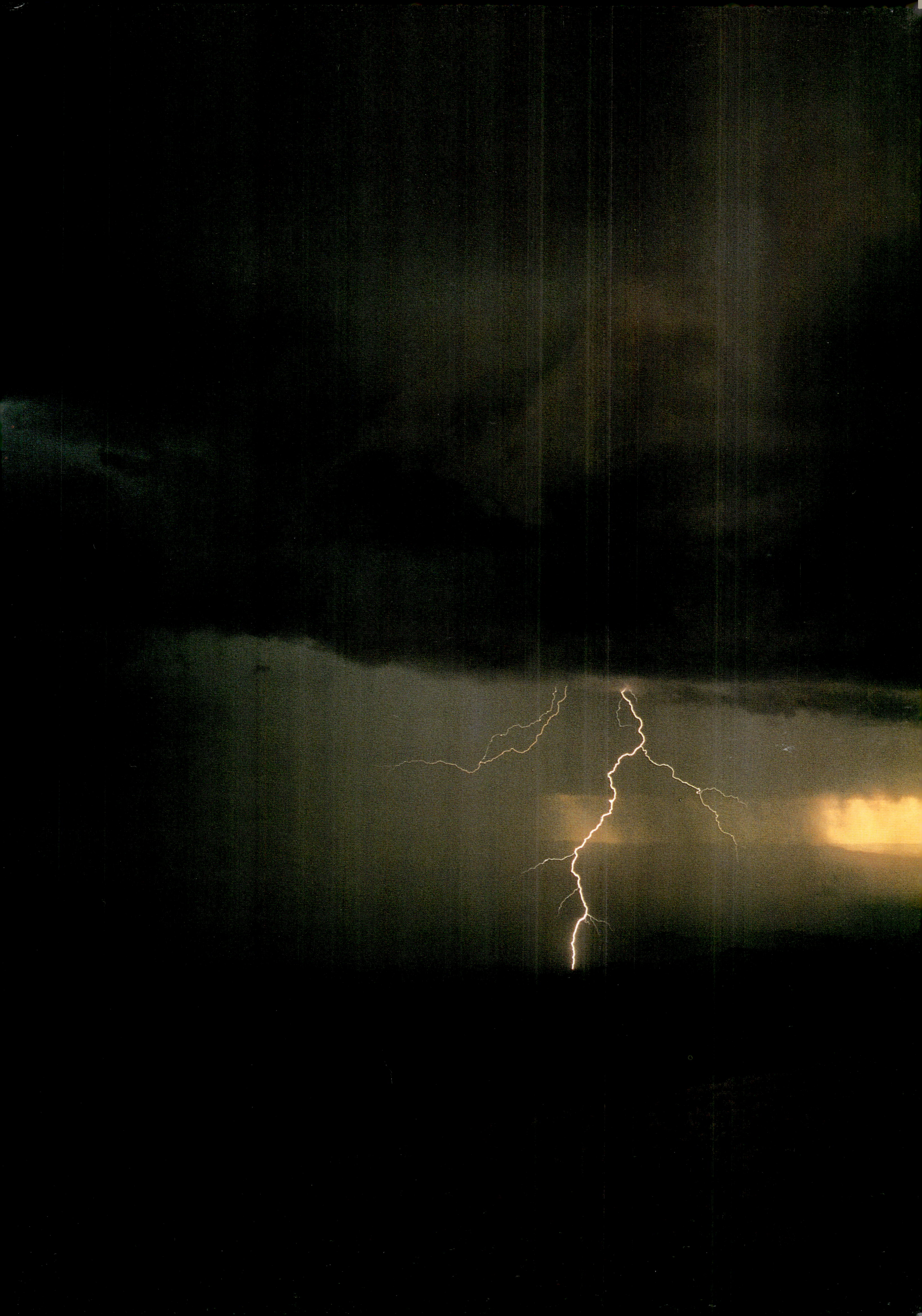

Today, earth watching from space plays a vital role in studies related to climate change, ocean currents, acid rain, deforestation, soil erosion, global ice patterns, the condition of rivers and lakes, crop diseases and pests, land-use patterns (including the encroachment of cities), and much more. These studies are helping us gain a better understanding of what is happening to the earth's atmosphere, its landmasses, and its oceans and rivers—and what we are doing to them.

HOLES IN THE OZONE

In the mid-1980s, data gathered by ground-based instruments and satellites in orbit around the earth conveyed findings about the chemistry of the atmosphere so startling and unexpected that scientists at first thought the data were in error. The observations showed evidence of a huge "hole" in the ozone layer over the Antarctic—a region of reduced ozone concentrations almost the size of the continental United States. The size of the hole varies with seasons and weather conditions, but depletions of from 30 to 50 percent have already been recorded, and they appear to be increasing from year to year. Similar, though less extensive, ozone holes have also been detected in the stratosphere over the Arctic.

Scientists are convinced the holes are being caused largely by industrial chemicals, primarily chlorofluorocarbons (CFCs), which are widely used as refrigerants and foam-blowing agents. These and other chemicals release chlorine and bromine compounds into the upper atmosphere, where they enter into repeating chain reactions in which they destroy thousands of ozone molecules without themselves being removed. A single chlorine atom can eat up as many as 100 thousand ozone molecules in this way.

It's been estimated that over the past few decades human activities have already reduced the global ozone layer by an average of about 3 percent. Scientists believe that the depletion is greater in the polar regions at certain times of the year because of the effects of wind vortexes in trapping the chemicals and the presence of ice clouds that promote ozone-destroying reactions.

Ozone is scattered throughout the earth's upper atmosphere, between about 15 and 36 kilometers in very low concentrations; if the entire layer were to be compressed to ground-level atmospheric pressure, it would be only 3 millimeters thick. This fragile barrier is all that stands between life on earth and the sun's biologically damaging ultraviolet radiation, known as UV-B. A decrease in ozone would result in an increase in the amount of UV radiation reaching the earth's surface. This in turn would cause an increase in the incidence of skin cancer in humans and damaging

effects on crops and marine ecosystems. (Phytoplankton, small organisms that live near the surface of the oceans and form the basis of the marine food chain, would likely be highly vulnerable to increased UV-B levels.) There is also some scientific evidence that increased ultraviolet levels could cause impairment of the immune system in animals and humans, which would lead to an increase in the incidence of infectious disease.

Governments around the world are attempting to phase out the use of chlorofluorocarbons and to find ozone-safe substitutes, but many environmentalists believe this is not being done fast enough. Perhaps the most disquieting aspect of the ozone problem is that CFCs are extremely long-lived. In fact, most of the CFCs we have already released are still in the lower atmosphere and haven't even begun to do their damage. "Even if we stop producing these compounds today, it will be decades, if not centuries, until the damage to atmospheric ozone ends," says atmospheric researcher Robert Watson of the National Aeronautics and Space Administration.

THE GREENHOUSE EFFECT: A PLANETARY THREAT?

The urgency to reduce global CFC emissions is also fueled by the fact that the chemicals are "greenhouse" gases. Along with carbon dioxide, water vapor, nitrogen oxides, and methane, they trap heat radiating from the earth's surface and prevent it from escaping into space, causing the earth's atmosphere to warm up. This greenhouse effect occurs naturally; in fact, without it, the earth's average temperature would be near $-18°$ C ($0°$ F) instead of $15°$ C ($59°$ F). On the other hand, too much warming can be just as inhospitable; a runaway greenhouse effect sent the surface temperature of Venus soaring to about $454°$ C ($850°$ F).

Since the beginning of the Industrial Revolution, and particularly since World War II, human activities have been adding greenhouse gases to the atmosphere at an increasing—and now alarming—rate. The level of carbon dioxide—mostly from the burning of carbon-containing fuels such as wood, coal, oil, and gas—has risen by about 25 percent since the middle of the nineteenth century. We are now pumping an estimated five to six billion tons of carbon dioxide into the atmosphere every year, and it's expected that greenhouse gases will double by the middle of next century. Scientific studies indicate that a doubling of carbon dioxide would increase the average temperature of the earth by about 1.5° to 4.5° C (3° to 8° F). If that doesn't sound like much, it should be remembered that the difference between the earth's current average annual temperature and that of the last full ice age is only about 5° C (9° F). It has been estimated that we're changing the climate ten to sixty times faster than normal, and

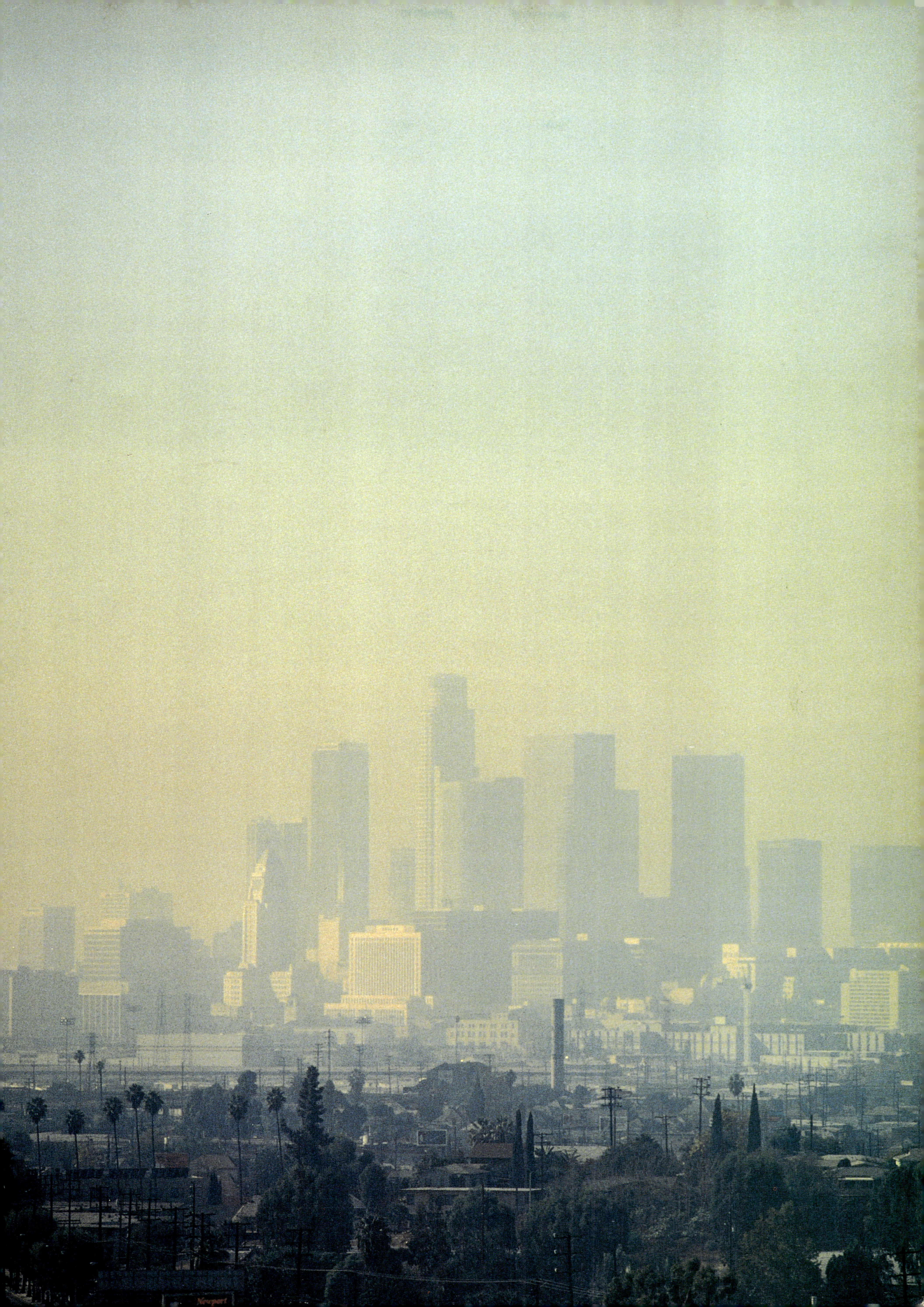

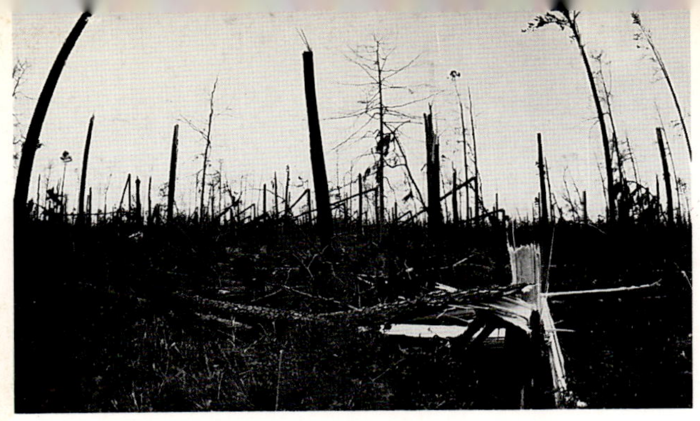 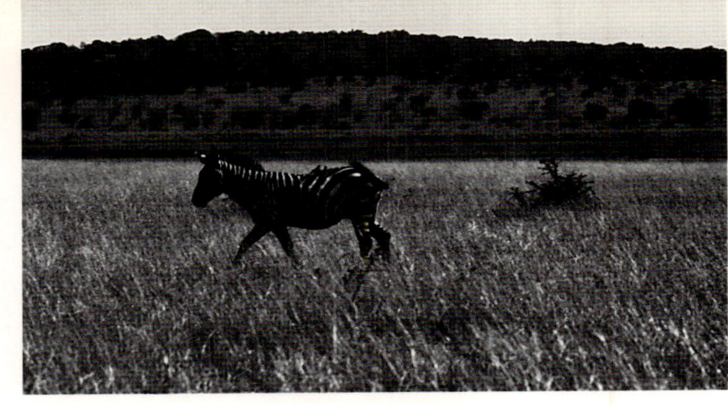

scientists say we will have to cut our annual emissions at least in half to stabilize atmospheric carbon dioxide at current levels—a task that will be difficult, given our economic and social dependence on fossil-fuel energy.

Although CFCs are at present far less abundant in the atmosphere than carbon dioxide, they have a greenhouse effect ten thousand times greater per molecule, and it's estimated they already account for about 15 percent of the greenhouse effect. There are a number of other greenhouse gases produced at least partly by human activities: methane (from agriculture and farming); nitrous oxide (from chemical fertilizers and the burning of fossil fuels and wood); and ground-level ozone (a major component of urban smog). The combined greenhouse effect of all of these chemicals may be nearly equal to that of carbon dioxide.

Has global warming caused by greenhouse gases from human activities already begun? Scientists can't yet say for sure; although some experts have expressed the opinion that it has, this is still controversial. Most climate experts, however, are convinced that warming is inevitable if atmospheric carbon dioxide levels continue to increase. Exactly how this warming will affect the climate is hard to predict—particularly on a regional or local basis—because the earth's atmosphere is so complex and still not very well understood. However, various studies suggest that normal temperature and rainfall patterns could be altered and there could be an increase in the variability of the weather. We may be in for bigger and more frequent swings between hot and cold, wet and dry, more powerful and damaging storms, more severe droughts—in short, more rude surprises.

One of the most significant effects may occur as a result of changes in cloud patterns. A changing climate could alter the delicate balance of their warming and cooling effects, with unknown but probably significant consequences.

Clouds are children of the relationship between the oceans and the atmosphere. Water evaporates from the ocean surface; rises into cool air; condenses, forming clouds; and eventually falls as rain. In this way, clouds act as couriers of heat and water around the globe. If a hotter climate were to cause the oceans to warm up, this would likely increase cloudiness and thus enhance both the warming and cooling effects of clouds. It might also change the location of certain types of clouds and alter global air circulation and rainfall patterns, causing a shift of desert and humid regions and possibly an intensification of storms.

In recent years, it appears we've been given a taste of what a more variable climate would be like. A number of extreme, record-setting weather events have occurred, and although it's impossible to prove they were caused by greenhouse warming (some scientists caution that they could merely be due to normal weather variability), they have nevertheless focused public attention on the possible consequences of climate warming. In February, 1990, a series of monster storms with winds up to 200 kilometers per hour battered Britain and other parts of Europe, causing widespread flooding, power outages, and property damage, disrupting transportation, and forcing the evacuation of thousands of people. In September, 1988, Hurricane Gilbert—the most powerful hurricane ever recorded in the Northern Hemisphere—pummeled the Caribbean, the Yucatán Peninsula of Mexico, and parts of the Texas coast, causing tens of billions of dollars in damage and leaving hundreds of thousands of people homeless. Tourists in Jamaica said they saw walls of water so high they couldn't see the sky. Hurricanes get their energy from warm surface waters over the ocean, and it was suggested that one of the reasons Gilbert was so ferocious was that the waters of the Gulf of Mexico were warmer than usual. Some scientists speculate that more powerful hurricanes might be one consequence of global warming.

During the summer of 1988, many regions of North America experienced a record-breaking weeks-long heat wave during which temperatures repeatedly soared above 40° C (104° F). Then, during the following winter, many parts of North America were hit by record-breaking cold temperatures and snowfalls (a development that led some to conclude, erroneously, that the greenhouse effect was just so much "hot air"). The summer of 1989 was also a terrible year for forest fires in the United States and Canada; in fact, in Canada it was the worst year in the eight decades that records have been kept.

On a global scale, the six warmest years on record occurred in the 1980s, and severe drought now regularly devastates agriculture in many regions of the world. Between 1986 and 1988, these droughts caused world grain reserves to drop to their lowest point in decades—"little more than is necessary to fill the pipeline from field to table," according to Lester Brown, president of the Worldwatch Institute. He notes that in 1989, for the first time, the United States did not produce enough grain to meet its own consumption needs.

Desertification is a matter for serious concern throughout much of Africa. Since the 1970s, the use of remote sensing data has enabled scientists to monitor the spreading of the deserts far better than they can on the ground, says Ted Maxwell, chairman of the Center for Earth and Planetary Studies at the National Air and Space Museum and a member of the *Blue Planet* advisory committee. He specializes in studying the movement of sand dunes in Africa, particularly in Egypt and Sudan. He has seen what he describes as "shallow ripples" in the "sand sheet"—the flat, open areas of sand where there are no dunes. Only a few centimeters

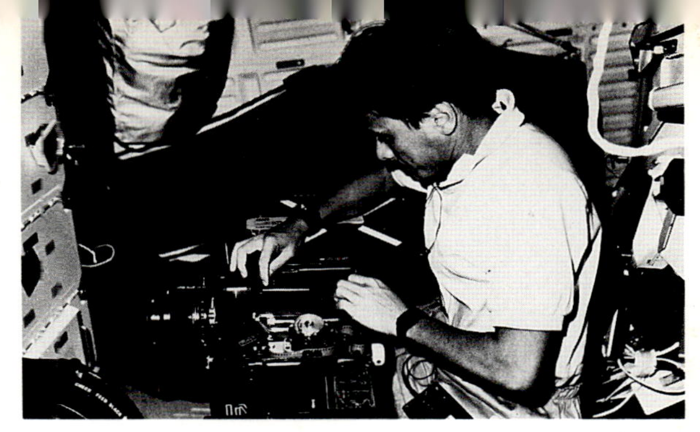

high, these ripples show up on satellite pictures as chevron-shaped patterns. "They're moving at a great speed—about a kilometer a year—and transporting a lot of sand. They're heading to the south and will eventually end up in the Sahelian zone." He notes that this phenomenon is not caused by human activities and it's impossible to predict how climate warming might affect it.

Space-based observations have been a great help to Maxwell; by overlaying satellite images taken in the 1970s with ones taken in the 1980s, he has been able to chart sand migration over much larger areas than would be possible on the ground. He says that it would take him months of planting stakes to measure sand dune movement on the ground, whereas analyzing satellite pictures takes only a couple of hours. Viewing the earth from space is a "fabulous tool" with which to learn about the human impact on the environment, he adds. "We're just beginning to use the data collected over the last twenty years to compare environmental changes due to natural factors and due to human impact. If you go out into the field, you'd never get a comprehensive view of it."

DISAPPEARING FORESTS

In trying to anticipate what will happen to the earth's climate it is not enough to focus solely on the role of the atmosphere. The oceans and landmasses—and especially the life they harbor—also play a crucial role. Plants in particular are key players; they not only remove carbon dioxide from the atmosphere and release oxygen into it, but they also lock up carbon in their tissues, making it temporarily inaccessible to the atmosphere. It has been estimated that we'd have to plant a forest half the size of the United States, or about one-third of all the land on earth now devoted to agriculture, to absorb the carbon dioxide put into the atmosphere by human activities.

Far from doing this, however, we're cutting down trees at a tremendous rate. All over the world, but particularly in the tropical rain forests, trees are being chopped down for lumber or burned to clear land for farming or cattle ranching. (Ironically, some of these farming operations fail anyway because tropical soils are unsuitable for crops.) Added to this are environmental stresses and pollution. The result is that forests everywhere are disappearing at an incredible pace—from the Himalayan foothills to the slopes of the Andes, from the Amazon to New Caledonia, from the west coast of North America to East Africa, from Central America to Australia. The United Nations World Resources Institute estimates that between 160,000 and 200,000 square kilometers per year of tropical rain forests are being lost—more than 2 percent of the

estimated 8 million square kilometers of forest remaining.

Forests in more developed parts of the world—in Europe and eastern United States and Canada, for example—have been logged for centuries and are now falling victim to acid deposition (including acid rain) caused by pollution from heavily industrialized regions. (The main sources of acid deposition are sulfur dioxide, produced mainly by industry, and nitrogen oxides, produced mainly by motor vehicles.) In Western Europe, vast areas of forest have been damaged by acid deposition; the Black Forest of Germany, for example, has been devastated. Central and Eastern Europe have even more serious problems.

Acid rain has also done tremendous damage to forests in northeastern North America. By the mid-1980s, trees in Ontario, Quebec, and New England were succumbing at a frightening rate, according to environmental scientists, and Quebec's maple sugar industry is now threatened by pollution. Studies have shown that damaged trees in Ontario had high levels of aluminum in their roots. It's believed that acid rain causes aluminum to be released from the soil.

In some regions, deforestation has contributed to droughts and the spreading of deserts, while in others it has created conditions leading to massive soil erosion and flooding because of the loss of trees that play a key role in anchoring topsoil and modulating water flows after rainfall. In India and Nepal, monsoon seasons have often turned into disasters because watersheds have been denuded of trees. Worldwatch Institute President Lester Brown has noted that deforestation leads to increased rainfall runoff and crop-destroying floods. Soil erosion is slowly undermining the productivity of one-third of the world's cropland.

The Amazon basin, which covers more than a third of the total land area of South America, contains the largest and one of the most threatened rain forests on earth. In 1990, Brazilian researchers estimated that about 400,000 square kilometers of the forest has already been destroyed, which amounts to at least 10 percent of the original forest cover. This figure is higher than any previous estimate and suggests that deforestation is occurring even faster than had been believed.

In Peru, more than 2,000 square kilometers of national forests and parks have been destroyed to make room for coca plantations to produce cocaine bound for U.S. and European drug markets. The coca plant has become the largest crop cultivated in the Peruvian Amazon; one region, the Upper Huallaga River Valley, generates more than $600 million in revenues a year. The coca plantations have caused not only deforestation and soil erosion but also severe pollution, as millions of tons of toxic chemicals used in the processing of cocaine are dumped into rivers and streams each year. Residues of fertilizers and pesticides also run off into

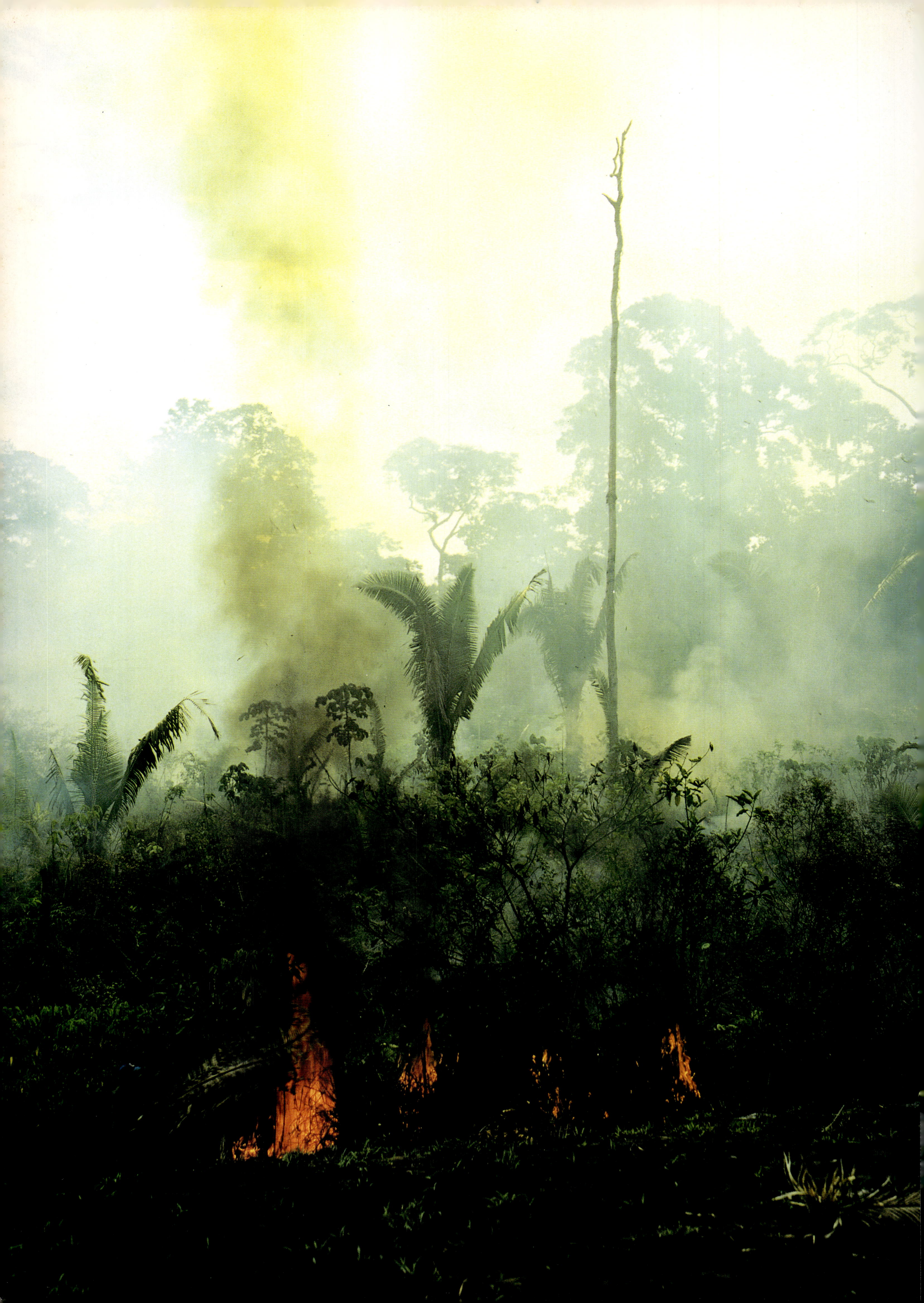

the water, and many species of fish, amphibians, reptiles, and crustaceans have already disappeared from Amazon waters. According to one study, coca production has directly or indirectly caused the destruction of 6,800 square kilometers of Amazon rain forest since the early 1970s.

The tremendous costs of deforestation are only beginning to be felt. For example, the Costa Rican government estimates that reforesting a watershed to prevent damage to a hydroelectric dam will cost twenty times what it would have cost to protect the original forest. But deforestation does not just kill off trees and cause soil erosion—it also destroys the plants and animals that form an integral part of intricate forest eco-systems. Although they cover only about 7 percent of the earth's surface, tropical rain forests are home to at least half and possibly up to 80 percent of all living species.

One of the richest of these biological treasure troves is Madagascar, a large island off the southeast coast of Africa that shelters a higher percentage of unique species of plants and animals than any other part of the world. Eighty percent of its eight thousand species of flowering plants exist nowhere else on earth; the same is true for half of the more than two hundred bird species and almost all of the reptiles, frogs, mammals, and the three thousand varieties of butterflies and moths. Today, only about 10 percent of Madagascar's forest cover remains, and this too is threatened.

Like many other threatened forest regions, Madagascar is believed to shelter large numbers of plant and animal species that we have not yet even identified. It is estimated that there may be between five and thirty million different species on earth, but scientists have catalogued only about 1.6 million, and many more millions may be killed off before we even know of their existence. The National Science Foundation has stated that the rate of extinction over the next few decades is likely to increase to at least one thousand times the normal rate and may ultimately result in the loss of a quarter or more of the species on earth. University of Pennsyl-vania biologist Daniel Janzen has compared this to burning down all the libraries in the world without finding out what's in them.

Scientists say that the demise of the tropical forests could cause a spasm of extinction unequaled since the disappearance of the dinosaurs. This biological destruction has consequences that will affect us in many ways. For example, it reduces the genetic richness and variety of life on earth. In fact, scientists have become so alarmed by this loss of biodiversity that they've established gene banks. This is a way of saving a species' genes even if the species itself is wiped out. Unfortunately, many species are becoming extinct before their genes can be preserved. Deforestation also causes the loss of plants and animals that could potentially be the source of important medicines. About a quarter of the pharmaceuticals used in the United States today were isolated from wild plants. The drug Capoten (captopril), which is used to treat high blood pressure, was isolated from the venom of the Brazilian pit viper. Janzen says he knows of three plants that might help in the treatment of AIDS—and they all grow in rain forests.

Preventing further destruction of the earth's forests will be difficult because much of the tree-cutting in developing countries is done in an effort to generate economic returns to reduce poverty and escape a crushing international debt burden. Some environmental groups are now promoting land-for-debt programs, which involve financial arrangements between developed and developing countries for forgiving debt in return for protection of rain forests and other wildlife areas. Between 1987 and 1990, there were fifteen debt-for-nature swaps sanctioned by the World Wildlife Fund in eight countries. A total of more than $97 million of debt was purchased for about $16.3 million—about 17 percent of the face value of the debt.

The intensifying demands on land and soils by human activities, particularly agriculture, are creating environmental problems. The United Nations Environment Program has estimated that soil erosion and spread-ing deserts are now claiming some 60,000 square kilometers per year—an area about twice the size of Belgium—threatening the livelihood of about 20 percent of the earth's population. In some areas, the loss of soil quantity and quality is so great that these practices have been described as "mining" or "quarrying"—that is, treating soil as a nonrenewable resource, rather than as a renewable one. For example, in Madagascar, the formerly lush cloud forests are scarred with barren strips from slash-and-burn clearing for agriculture and overgrazing by cattle, which, in only a few years, result in the loss of the thin layer of topsoil. The World Wildlife Fund says the island is the most eroded region on earth. From space, we can see rivers all over the world turned brown by silt, disgorging tons of valuable topsoil into the oceans.

RIVERS, LAKES, AND OCEANS

The earth's water resources are also under unending assault from human activities. Thousands of tons of toxic chemicals and human waste and sewage are dumped into rivers, lakes, and oceans each year. In the St. Lawrence River in Quebec, only about 400 beluga whales remain; corpses of this endangered species wash up on the shore so contaminated with industrial chemicals and heavy metals that they must be handled as toxic waste. The 2,500-kilometer-long Ganges River, one of the most polluted rivers in the world, receives the raw sewage of nearly one-third of India's

850 million people and the chemical leavings of dozens of industries that have no waste treatment facilities at all. It has been estimated that bacterial contamination of Indian rivers is responsible for two-thirds of the illness suffered by Indians, most of it affecting children.

The Great Lakes, which provide water to forty million North Americans, are polluted with hundreds of toxic chemicals and other air- and water-borne pollutants that are already affecting animal life in and around the lakes. People are often advised to cut down on eating fish caught in the lakes; according to one study, eating contaminated fish may cause "subtle damage" to the brain and nervous system.

Tens of thousands of lakes in eastern North America and Europe are being acidified by rain containing sulfur and nitrogen pollutants from industry and the use of motor vehicles. In some regions, the acid content of rain can be 100 times higher than unpolluted rain. Acidification is also affecting groundwater and soils, and scientists have found that it makes toxic metals more mobile; for example, fish are likely to pick up more mercury in acidified lakes.

The oceans too are showing ominous signs of being overwhelmed by pollution from oil spills, toxic chemical discharges, raw sewage, and the dumping of hazardous medical wastes and tons of plastic. During the summer of 1988, people vacationing on Atlantic beaches off New York and New Jersey were aghast to find AIDS-contaminated blood vials and used hypodermic needles washing up on shore. The year before, U.S. eastern shores were littered with dying dolphins. Beaches all over the world are routinely strewn with plastic spit up by oceans now choking on the stuff. The plastic is killing sea life. Birds die because their beaks have been locked shut by the plastic rings used to carry six-packs or because they've swallowed plastic beads. Giant sea turtles have died from ingesting plastic bags they've mistaken for jellyfish. Seals and whales have also fallen victim to plastic pollution. All of this has happened in just thirty years, according to an oceanographer who easily collects thousands of pieces of plastic, even far from shore, by dragging nets behind a boat.

Like the atmosphere and the biosphere, the oceans play a central role in regulating the earth's climate. Transporting heat around the planet is one of the major ways they do this, and they can have a dramatic effect. For example, during the winter of 1982–83, severe global weather attributed to El Niño—a warming of the surface waters of the Pacific Ocean off the coast of South America—caused more than $8 billion in damage. The 1982–83 El Niño was the strongest in at least a century. Another, in 1987–88, was not as strong, but scientists believe it helped set the stage for the severe heat and drought that occurred in North America during the summer of 1988. El Niño's cousin, La Niña, which is characterized by a cooling of surface waters, also has far-reaching climatic effects.

The oceans also affect the climate by acting as a major repository for carbon. In fact, one of the great uncertainties in predicting how much climate warming will result from our pumping of carbon dioxide into the atmosphere is the extent to which extra carbon will be absorbed and rendered inaccessible in the oceans.

If major climatic changes occur, the oceans are likely to be themselves significantly affected. There would probably be a rise in sea level, partly because warmer waters would expand in volume and partly because climate warming might cause melting of polar ice caps and glaciers. It's been estimated that the sea level has risen about 10 to 15 centimeters during the past century, and one scientific study gives a "best guess" figure of another 7 to 25 centimeters by the year 2030. Some projections suggest it is possible that the sea level could rise as much as two meters, which would have a devastating effect in many coastal regions of the world. Some shorelines and islands might disappear altogether, displacing millions of people. Sea level rise could have far-reaching social and economic effects in low-lying coastal areas, one study notes, pointing out that a one-meter rise in sea level would eliminate 15 percent of Bangladesh, affecting some ten million people. In 1987 and 1988, Bangladesh suffered massive death and destruction from severe flooding, tidal waves, and cyclones. Likewise, in Egypt millions of people living on the arable land of the Nile River and its delta are vulnerable to the effects of drought, flooding, and rising sea levels.

On the other hand, water levels in inland lakes, such as the Great Lakes, might drop significantly—perhaps as much as two meters—because of increased evaporation in a warmer climate. This could cause water pollutants and toxic materials to become more concentrated and therefore more dangerous. It could also lead to the destruction of wetlands, which provide habitat for many species. And there would be an economic impact on tourism, energy production, agriculture, shipping, and fresh water supplies for cities.

In light of these problems, it's easy to see why studying the oceans has taken on a new urgency, but it's a challenging task because they are so vast and inaccessible. Although we have used the oceans as a source of food and transportation for millennia, they are, in many respects, as untouched a frontier as space. When we venture into the ocean depths, we must again carry our life-support system with us—vehicles that protect us from the near-freezing cold, crushing pressures, and lack of air. But unlike space, the ocean depths have rewarded us with the discovery of strange and unsuspected life forms living where water warmed by submarine volcanic vents streams out of fissures in the ocean floor. Amazingly, these

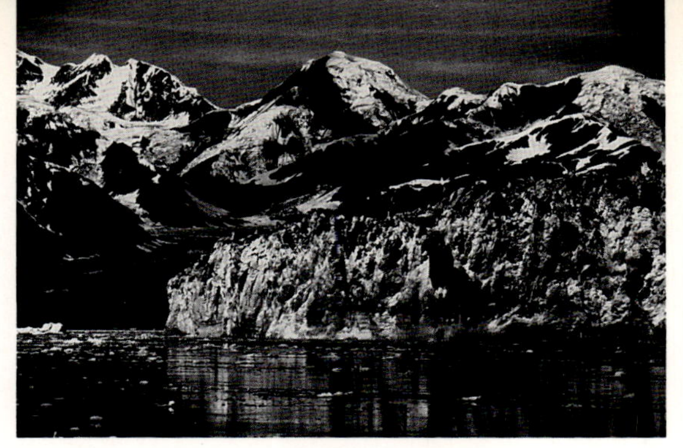

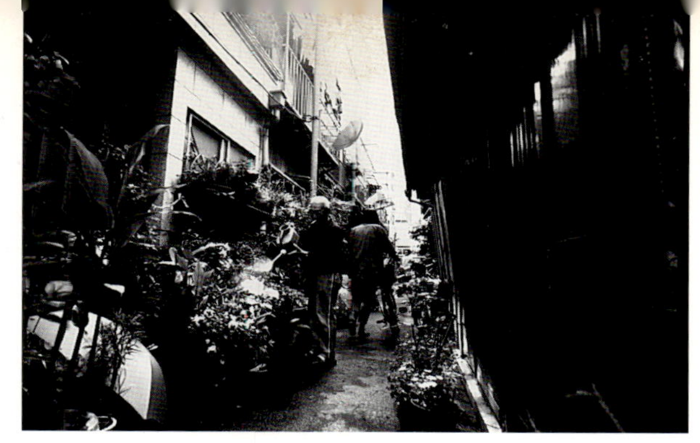

exotic creatures survive without sunlight and in a highly acidic environment, aided by bacteria that produce food from the normally poisonous chemicals spewed out through the vents.

As for studying the surface of the oceans, the task has been made much easier by the advent of satellites, which provide a frequency and extent of coverage that scientists cannot hope to match by using ships or even unmanned buoys. "Satellites provide a global snapshot of the ocean," says D. James Baker, president of Joint Oceanographic Institutions Inc. and a member of the *Blue Planet* advisory board. "Before satellites, all we had were ships or instruments that sat in one place or drifted around. To see all of the ocean at once was more expensive than anyone could afford, so most of it was not being monitored."

He says space observations have helped scientists to obtain better measurements of surface waves, winds, and currents, all of which are important in understanding the way the oceans redistribute heat around the globe.

Satellite data have also helped researchers studying life in the oceans. Space images first revealed the unexpected extent of chlorophyll blooms in the North Atlantic, which provided a way of assessing the distribution of the small plants called phytoplankton that form the basis of the marine food chain. "These blooms extend over thousands of miles for a couple of weeks in the spring," says Francis Bretherton. "We never realized that this was happening simultaneously over such a wide area. It completely changed our view about how these plants play a role in the uptake of carbon in the ocean."

Observations by humans have also played a role in increasing our knowledge of the oceans. On a shuttle flight in 1984, astronaut Paul Scully-Power discovered previously unseen currents in the ocean by taking pictures of surface waves with the sun glinting off them at just the right angle. "These pictures showed there is far more structure in the surface of the ocean than we dreamed," says Bretherton. "If he hadn't pointed his camera and brought back those fascinating pictures, we would never have realized that was going on."

earth. If, in some distant future, space adventurers from afar venture upon our planet, they will find much evidence of life—and perhaps too much evidence of dying.

There have been many massive extinctions of species in the past—about a dozen before human beings even walked the earth. The dinosaurs, which were a highly successful species for 100 million years—far longer than humans, who have been around for only about two million years—may have been done in by rapid climate change, possibly caused by the impact of a large meteorite perhaps 10 kilometers in diameter. Interestingly, scientists have calculated that the impact would cause short-term global cooling by injecting large amounts of dust into the stratosphere, followed by a long-term greenhouse warming due to the release of carbon dioxide from the oceans.

The important point, however, is that such a sudden, massive change in the climate—whatever the direction of the change—would leave plants and animals struggling to adapt to a totally new habitat and many would inevitably succumb to extinction. It is the pace of change that is perhaps the greatest threat to species survival. It's disquieting, therefore, that some scientists believe we are changing the earth's climate at a rate unprecedented in the last million years. They note that, if humans suffer the same fate as the dinosaurs, we will have the dubious distinction of being the first species to have brought extinction upon ourselves.

And yet, there is cause for hope in the fact that we have recognized the threat to the well-being of our species and our planet, and that we're struggling to understand the nature and dimensions of that threat. The first step in solving a problem is acknowledging that it exists. The vivid images of earth from space have helped us to do this—scientifically, philosophically, and emotionally. They've done more than merely alert us to the dilemmas we face in trying to preserve our global life-support system—they have also served to remind us, in ways we have never before imagined, just why the earth is so worth saving.

Human beings haven't ventured very far into space, but where we've gone we've looked for signs of life, any kind of life. So far, our quest has been fruitless. Neither the Apollo missions nor the Viking landers found any evidence of life on the moon or Mars—not even microbes in the soil. Mars is cold and lifeless; Venus, unbearably hot and just as lifeless. Between them, in our solar system's narrow zone of habitability, is the oasis we call

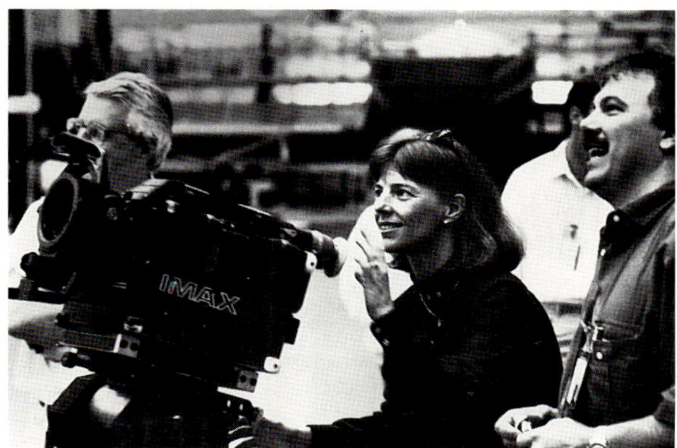

The spacefarer approaching earth sees it
first as a dazzling blue-and-white planet—
testament to the dominance of its oceans
and atmosphere. The brilliance of the planet
provides a stark contrast with the utter
blackness of space and forces upon us a
disquieting awareness of earth as a haven of
life in a universe that is mostly cold and
lifeless. These images forced us to realize as
never before that earth has boundaries, that
it is alone, and that it is all we have.

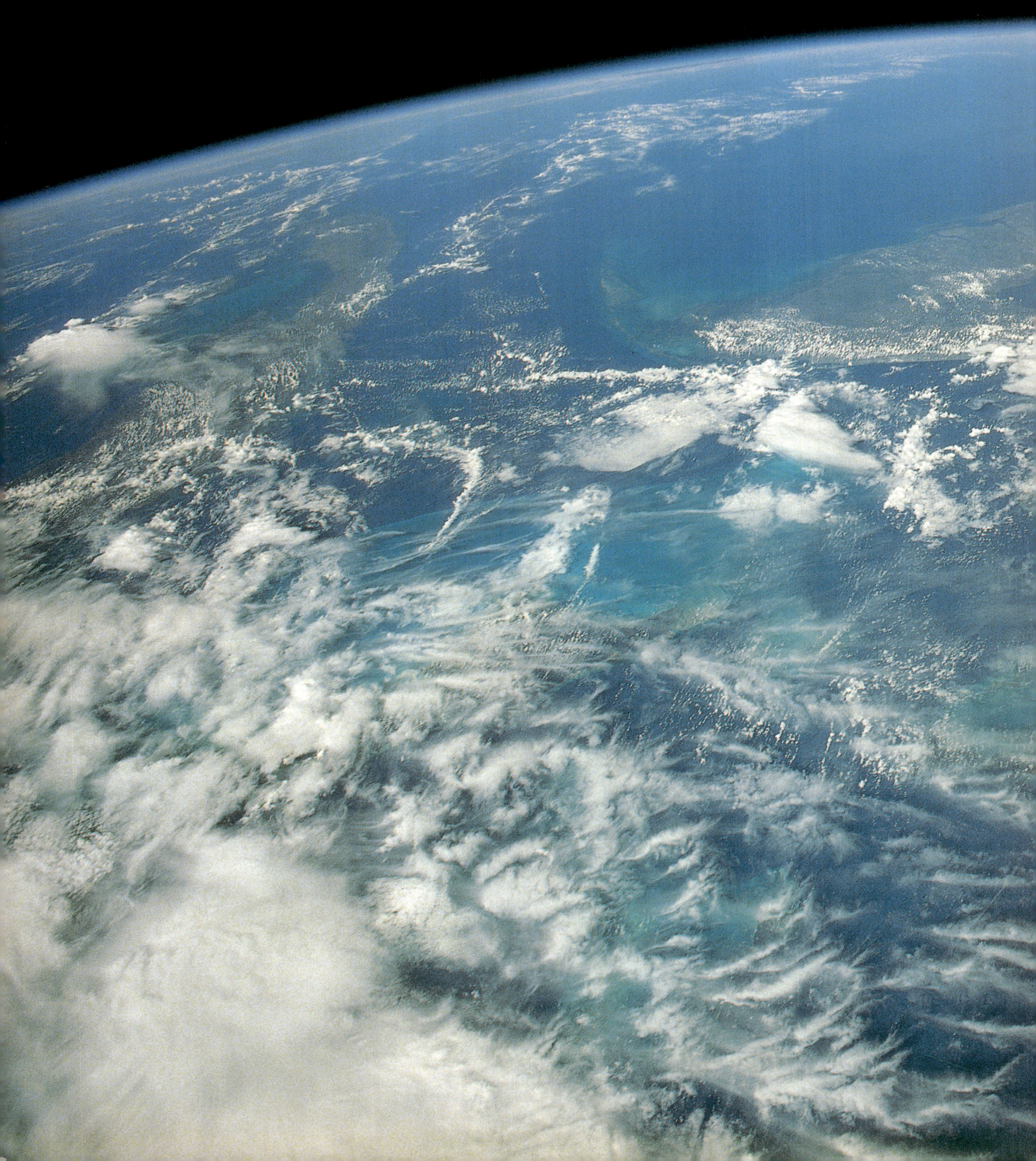

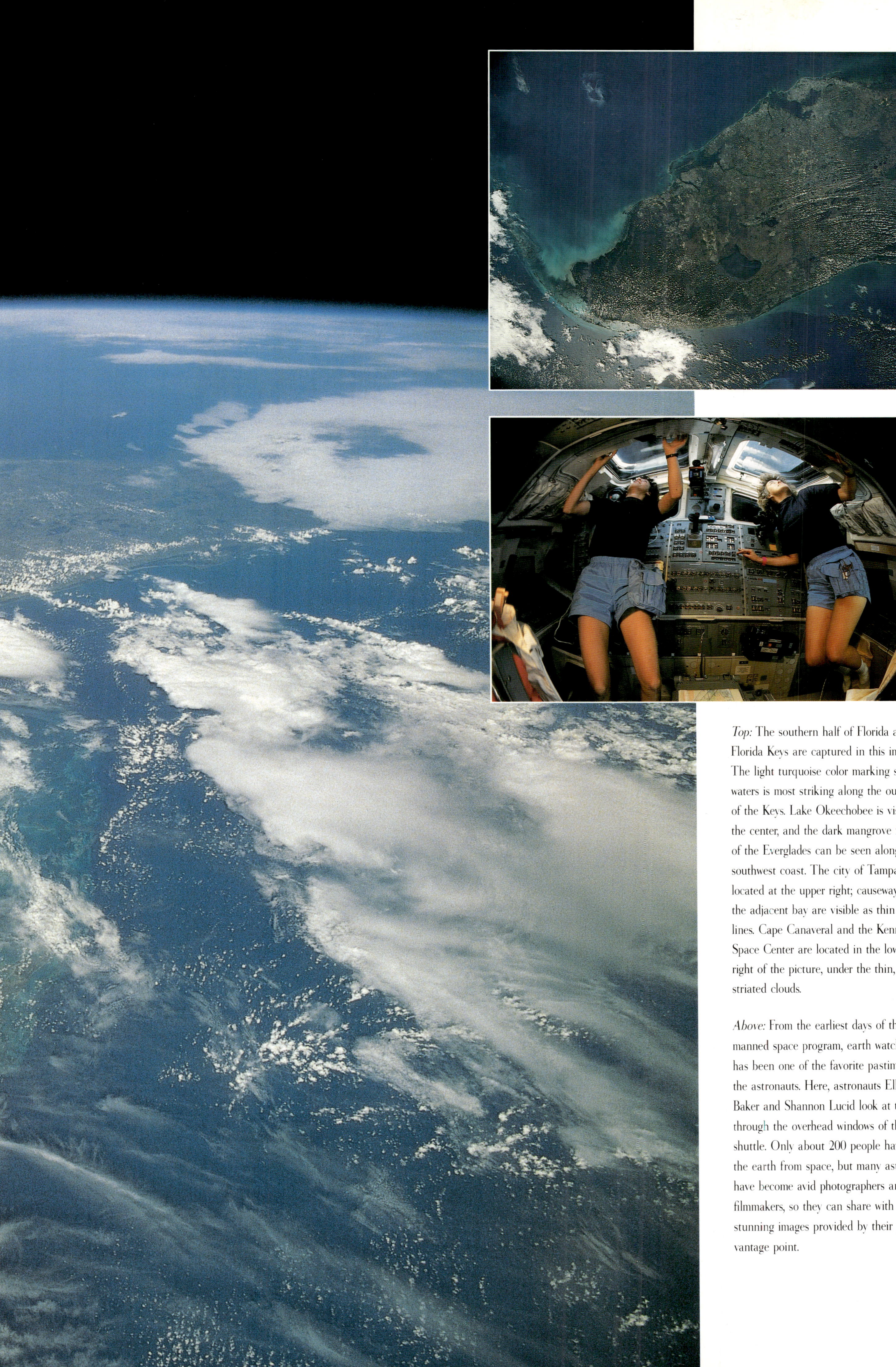

Top: The southern half of Florida and the Florida Keys are captured in this image. The light turquoise color marking shallow waters is most striking along the outer edge of the Keys. Lake Okeechobee is visible in the center, and the dark mangrove forest of the Everglades can be seen along the southwest coast. The city of Tampa is located at the upper right; causeways across the adjacent bay are visible as thin white lines. Cape Canaveral and the Kennedy Space Center are located in the lower right of the picture, under the thin, striated clouds.

Above: From the earliest days of the manned space program, earth watching has been one of the favorite pastimes of the astronauts. Here, astronauts Ellen Baker and Shannon Lucid look at the earth through the overhead windows of the space shuttle. Only about 200 people have seen the earth from space, but many astronauts have become avid photographers and filmmakers, so they can share with us the stunning images provided by their privileged vantage point.

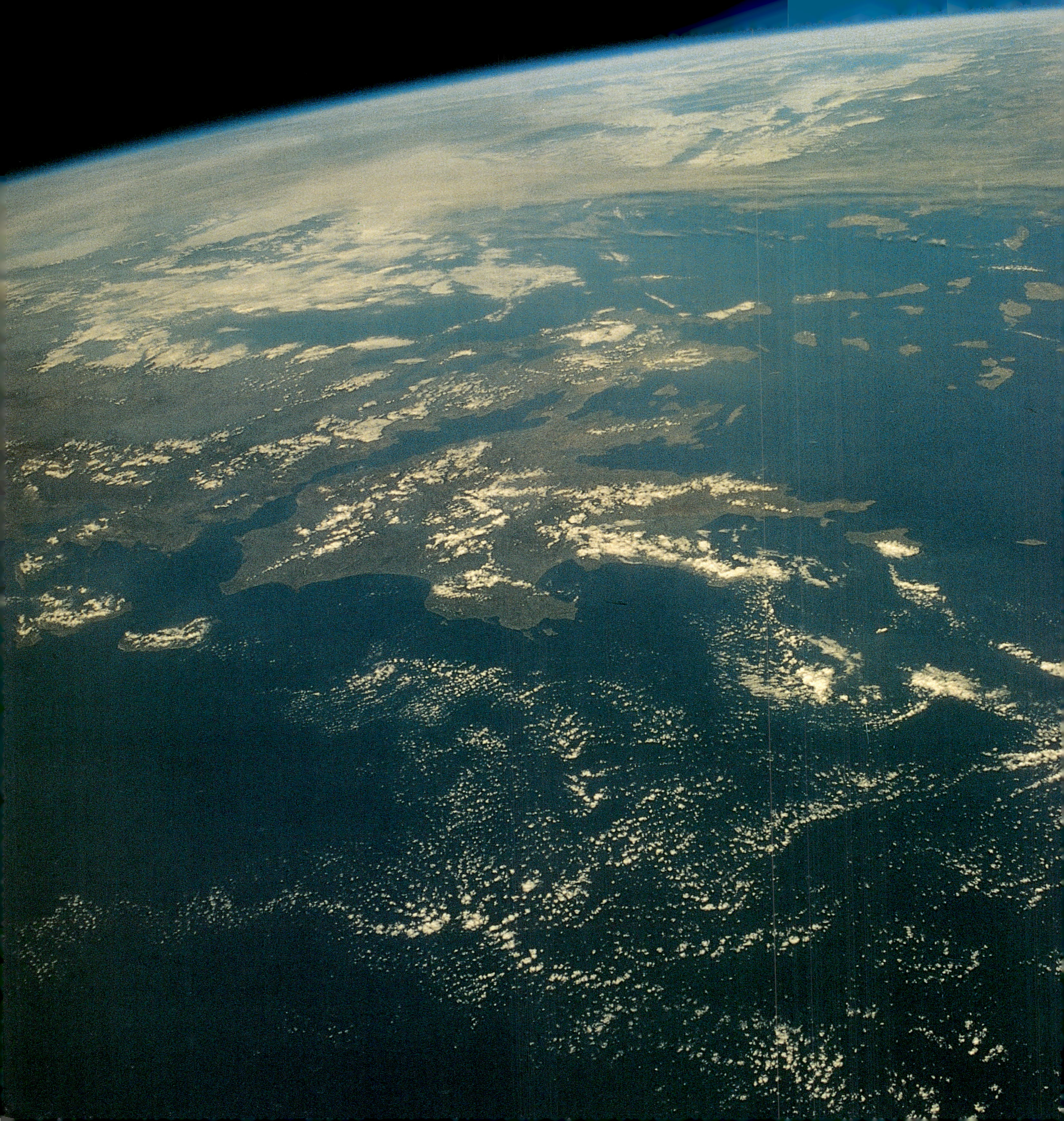

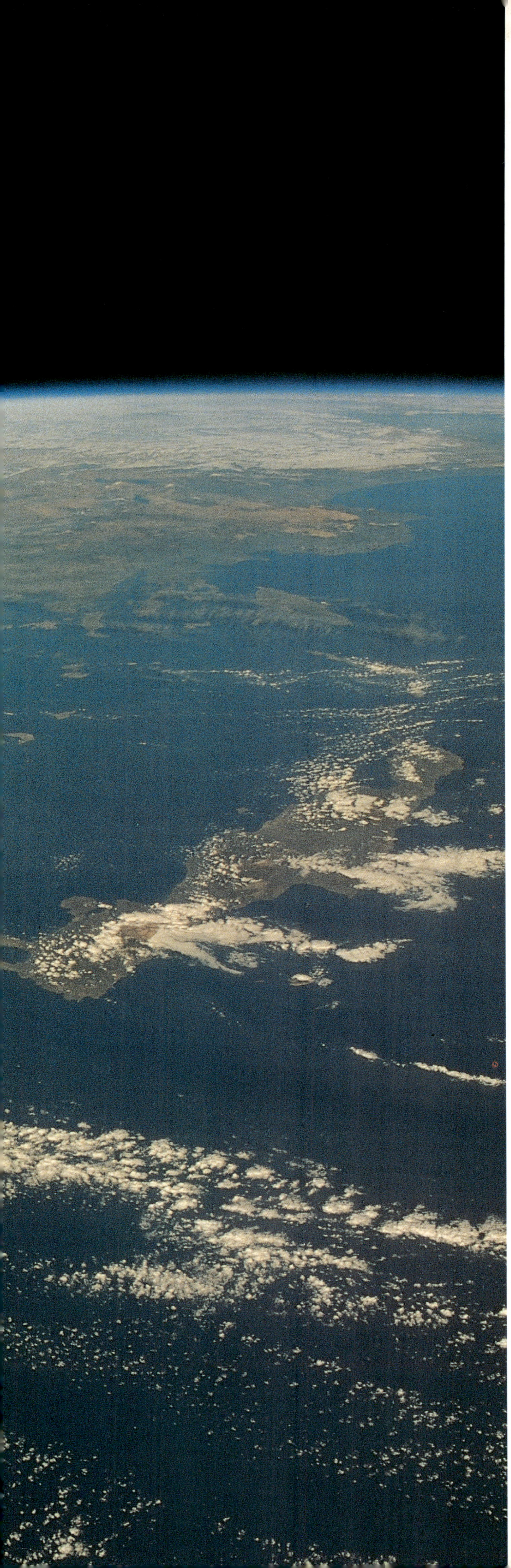

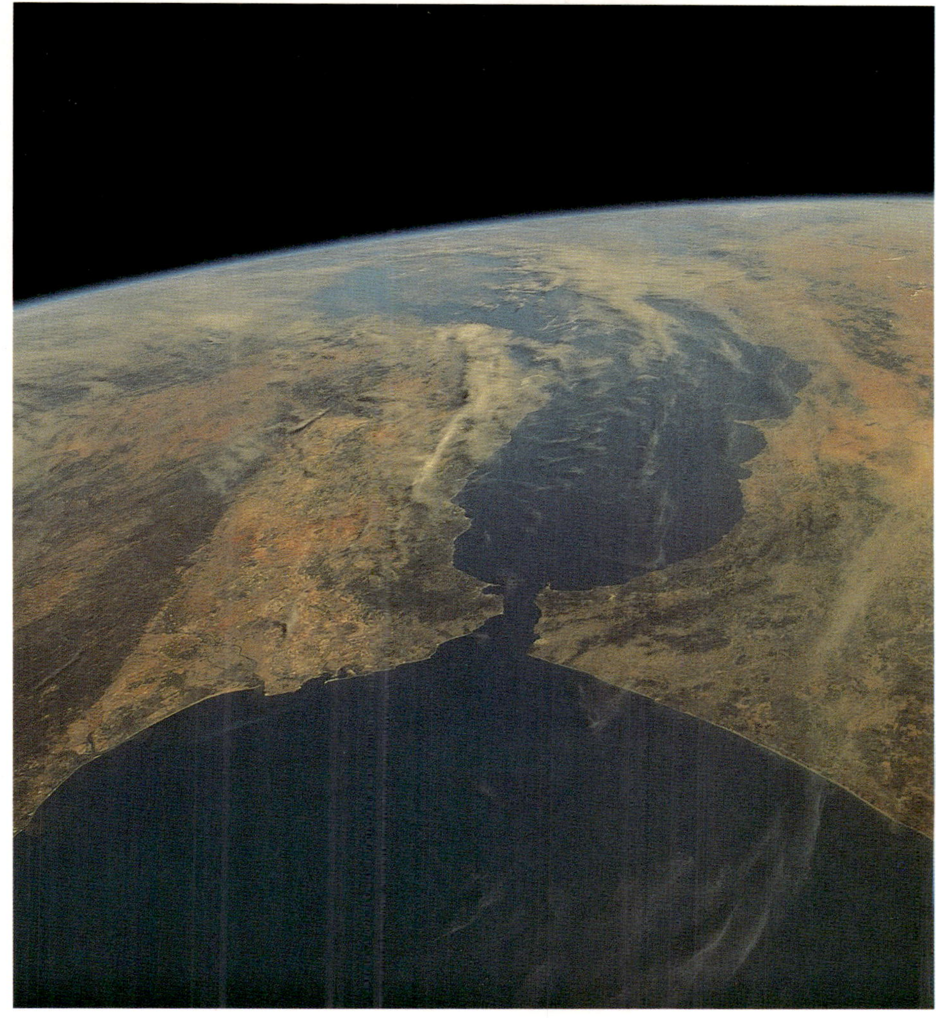

The Mediterranean is one of the most impressive sights from space. In the center right of this picture is the island of Crete, the site of the ancient Minoan civilization, the first in Europe. The small ring-shaped island north (left) of Crete is Santorini, the remnant of a huge volcano that erupted about 1628 B.C. Greece and the Greek Islands can be seen to the north and the coast of Turkey can be seen across the Aegean Sea.

The Strait of Gibraltar—only 13 kilometers wide—separates Morocco from Spain and links the Atlantic Ocean with the Mediterranean Sea. The Atlas Mountains, which arc across Morocco from the Atlantic to the Mediterranean, contain some of the highest peaks in North Africa. Clouds form over the mountains when moisture-laden air, forced upward, cools and condenses.

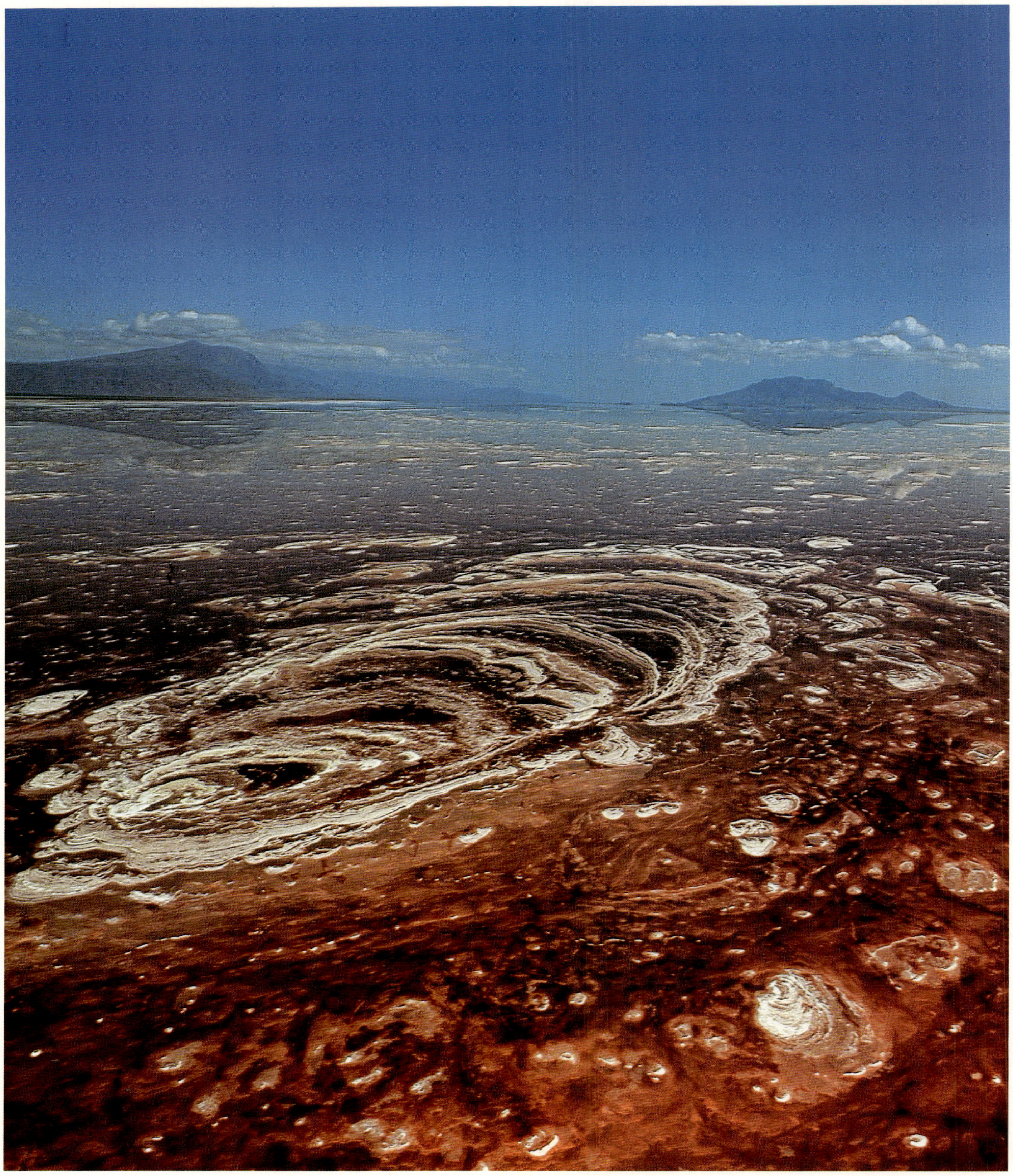

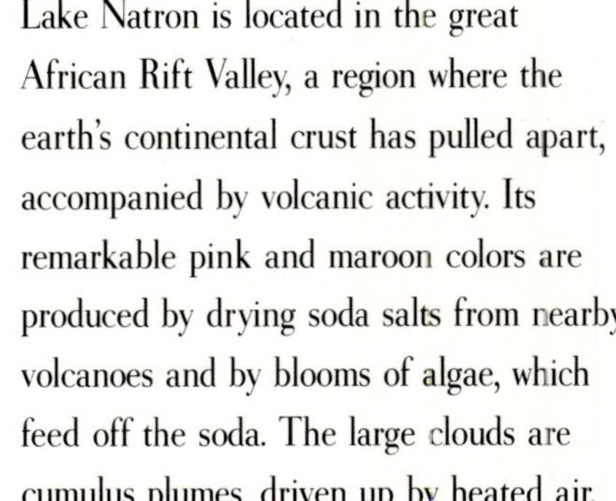

Lake Natron is located in the great
African Rift Valley, a region where the
earth's continental crust has pulled apart,
accompanied by volcanic activity. Its
remarkable pink and maroon colors are
produced by drying soda salts from nearby
volcanoes and by blooms of algae, which
feed off the soda. The large clouds are
cumulus plumes, driven up by heated air.

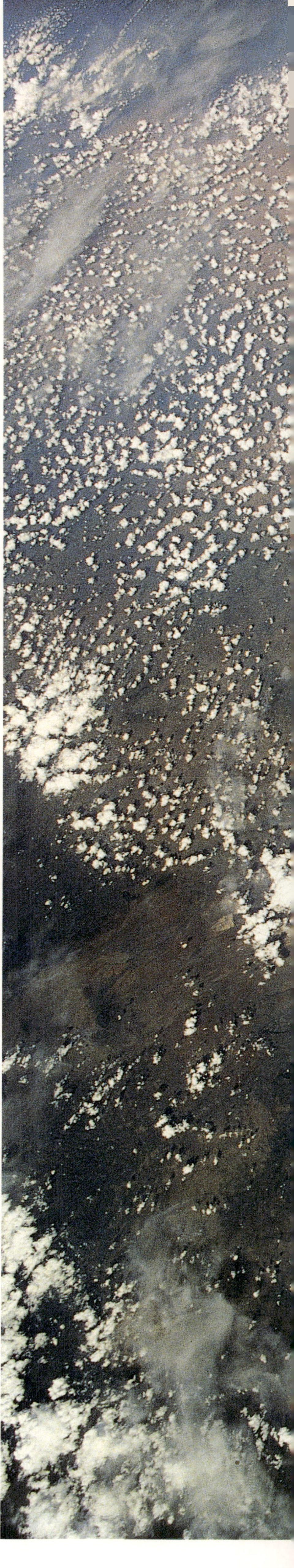

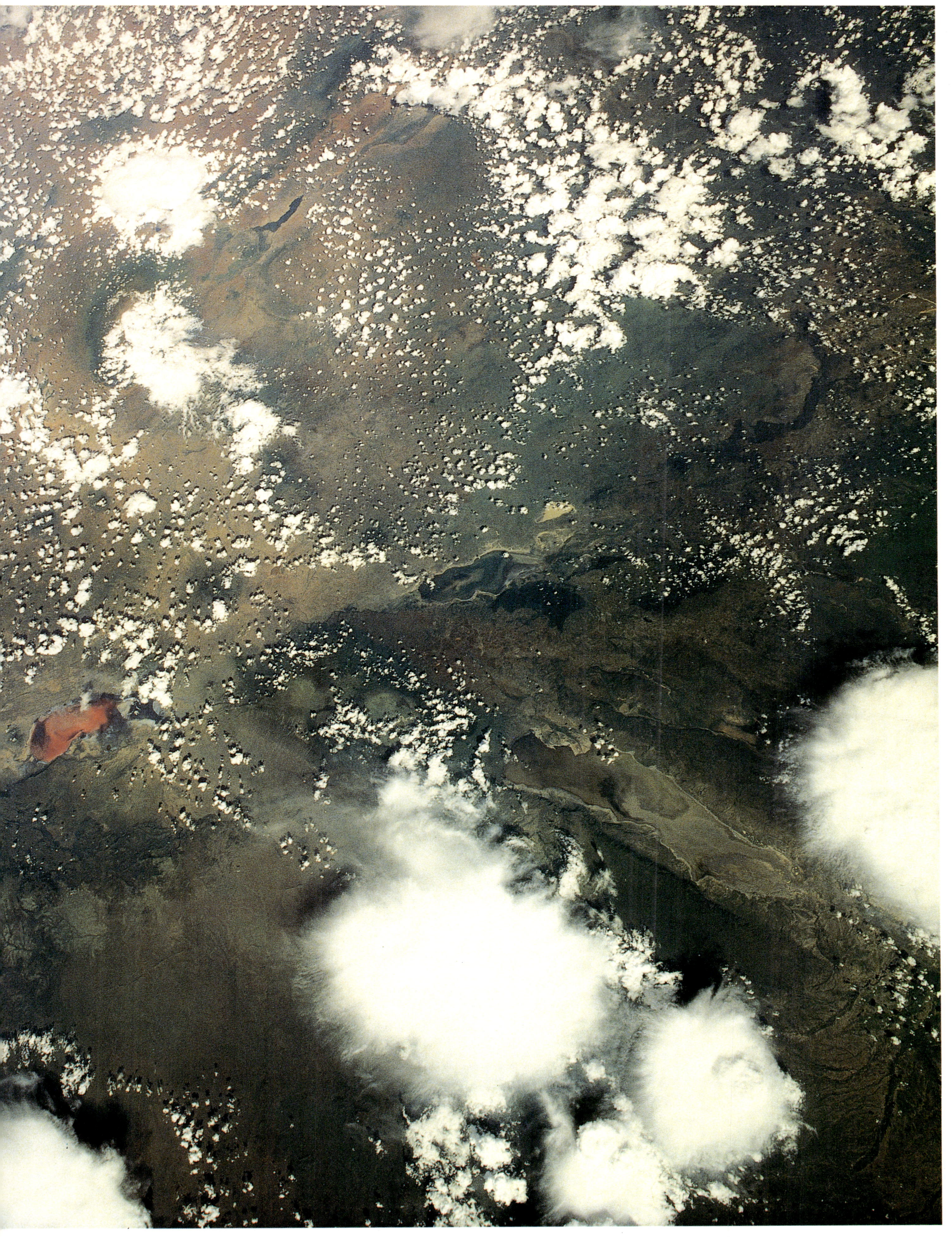

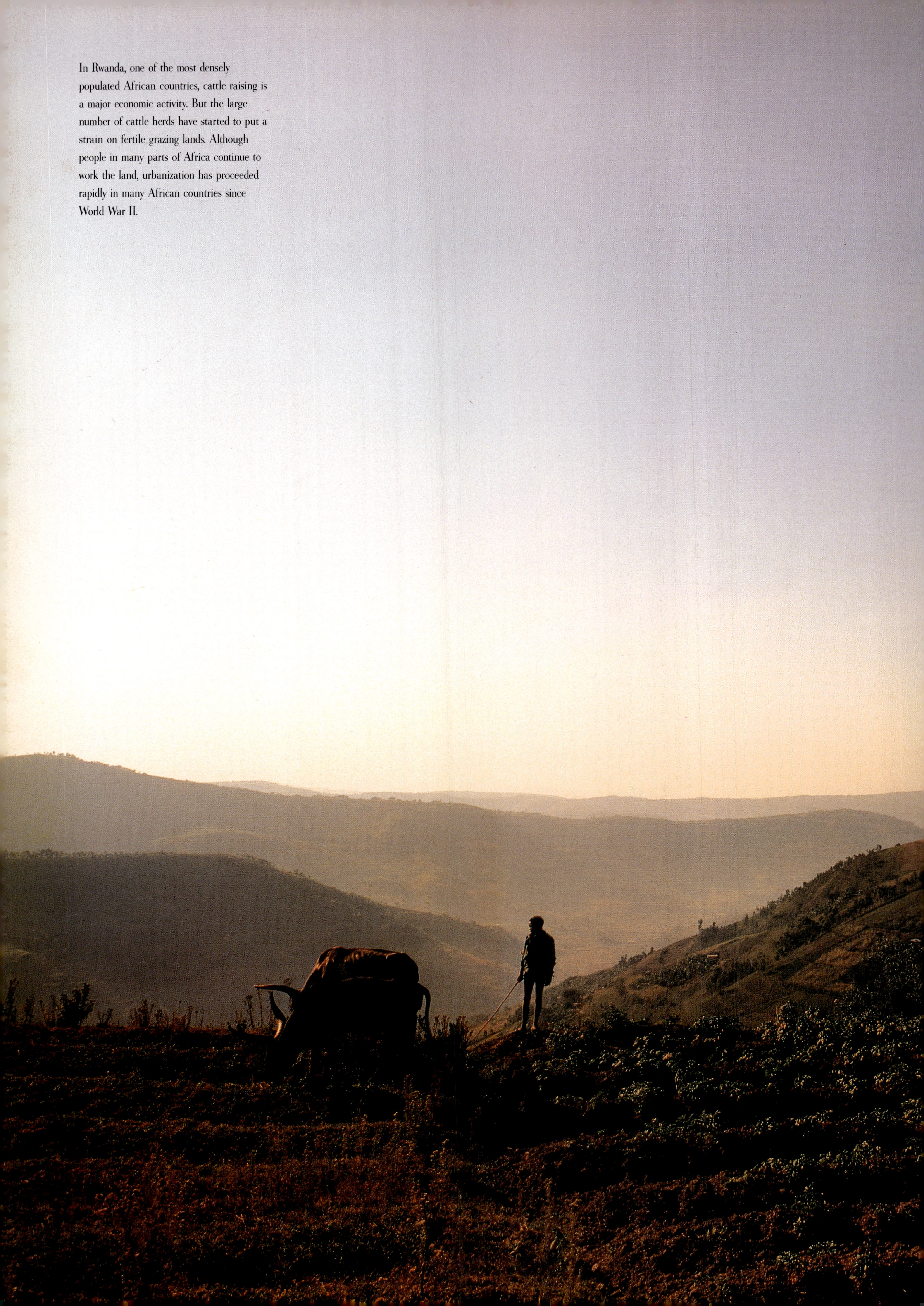

In Rwanda, one of the most densely
populated African countries, cattle raising is
a major economic activity. But the large
number of cattle herds have started to put a
strain on fertile grazing lands. Although
people in many parts of Africa continue to
work the land, urbanization has proceeded
rapidly in many African countries since
World War II.

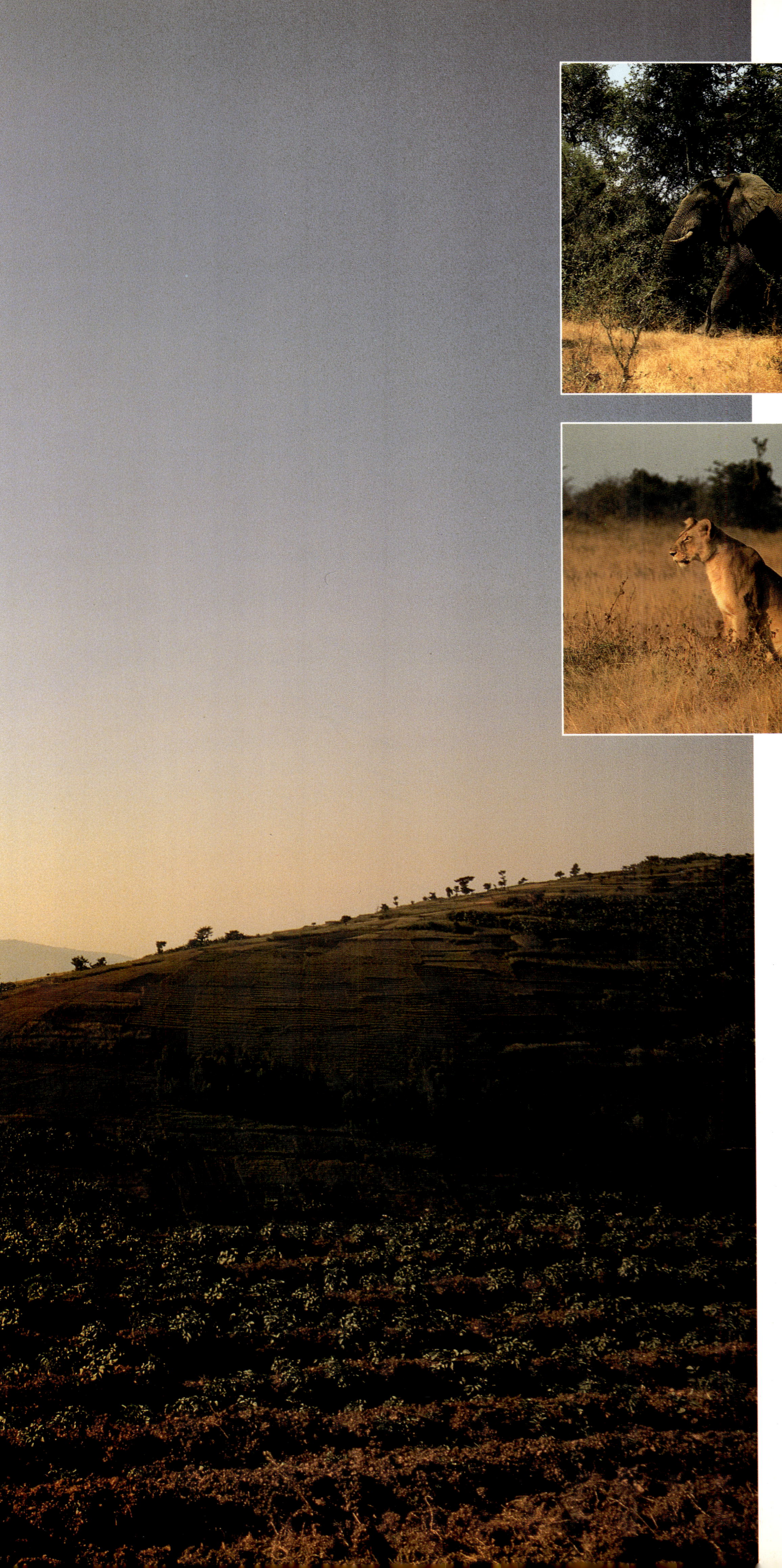

Top: The largest surviving land animal and the second-largest mammal on earth (after the whale), the elephant is in danger of extinction. Of the hundreds of elephant species that have existed throughout the earth's history, only two remain—the Asian elephant and the African elephant. The elephant seen here is roaming through Parc National de L'Akagera in the African country of Rwanda. Despite efforts to protect them, elephants continue to be illegally hunted for their tusks, the major source of commercial ivory. Poachers have left elephants bleeding and half-dead after cutting off their tusks with chain saws. Conservationists have endeavored to put an end to poaching by seeking to ban the international trade in ivory.

For most of the approximately two million years human beings have been on earth, we have lived in relative harmony with nature and with the plants and animals that share the planet. But human population growth is now threatening this balance. Only in the past few decades have we come to appreciate that our activities can strain the planet's life-support system to its limits.

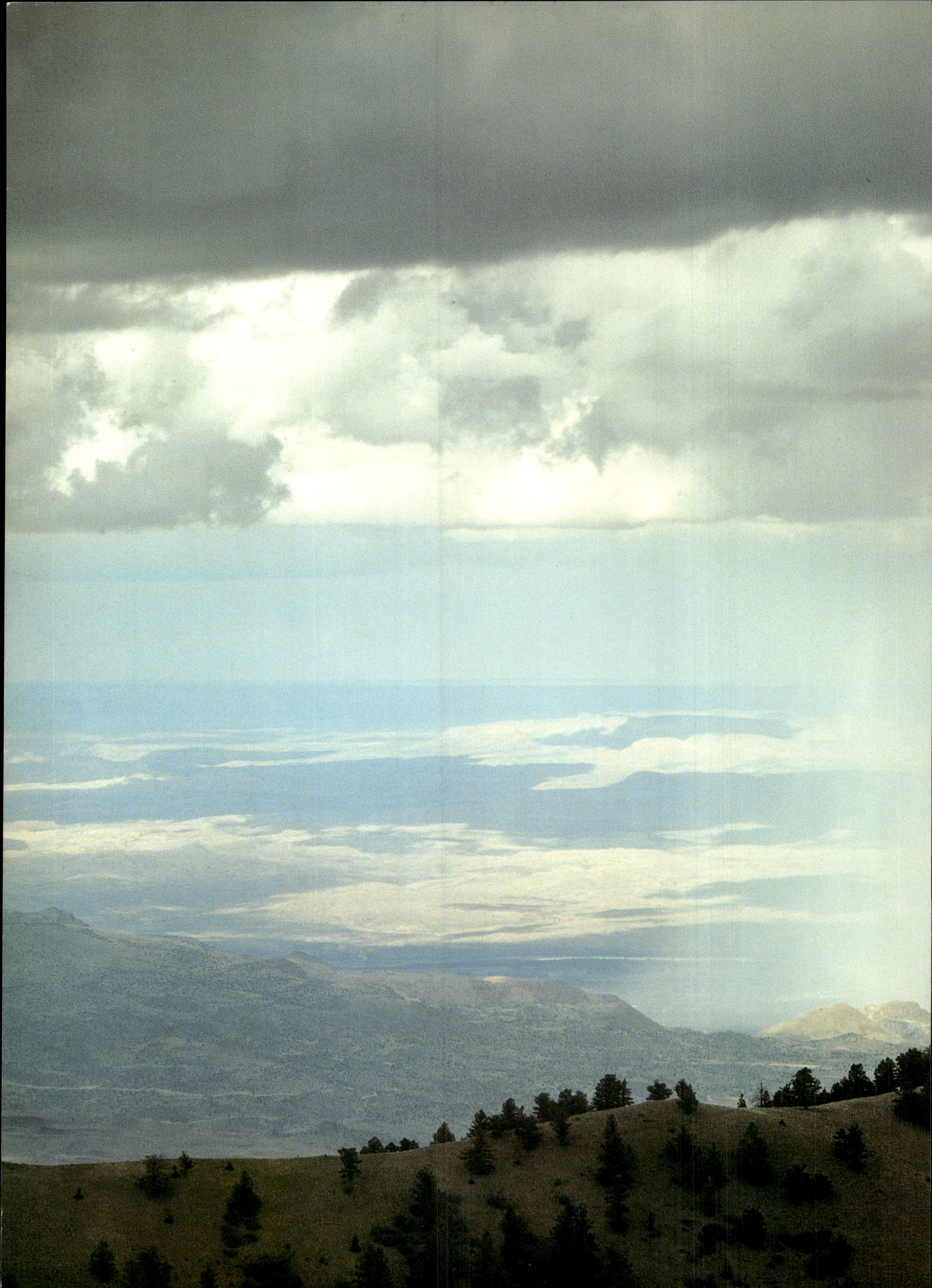

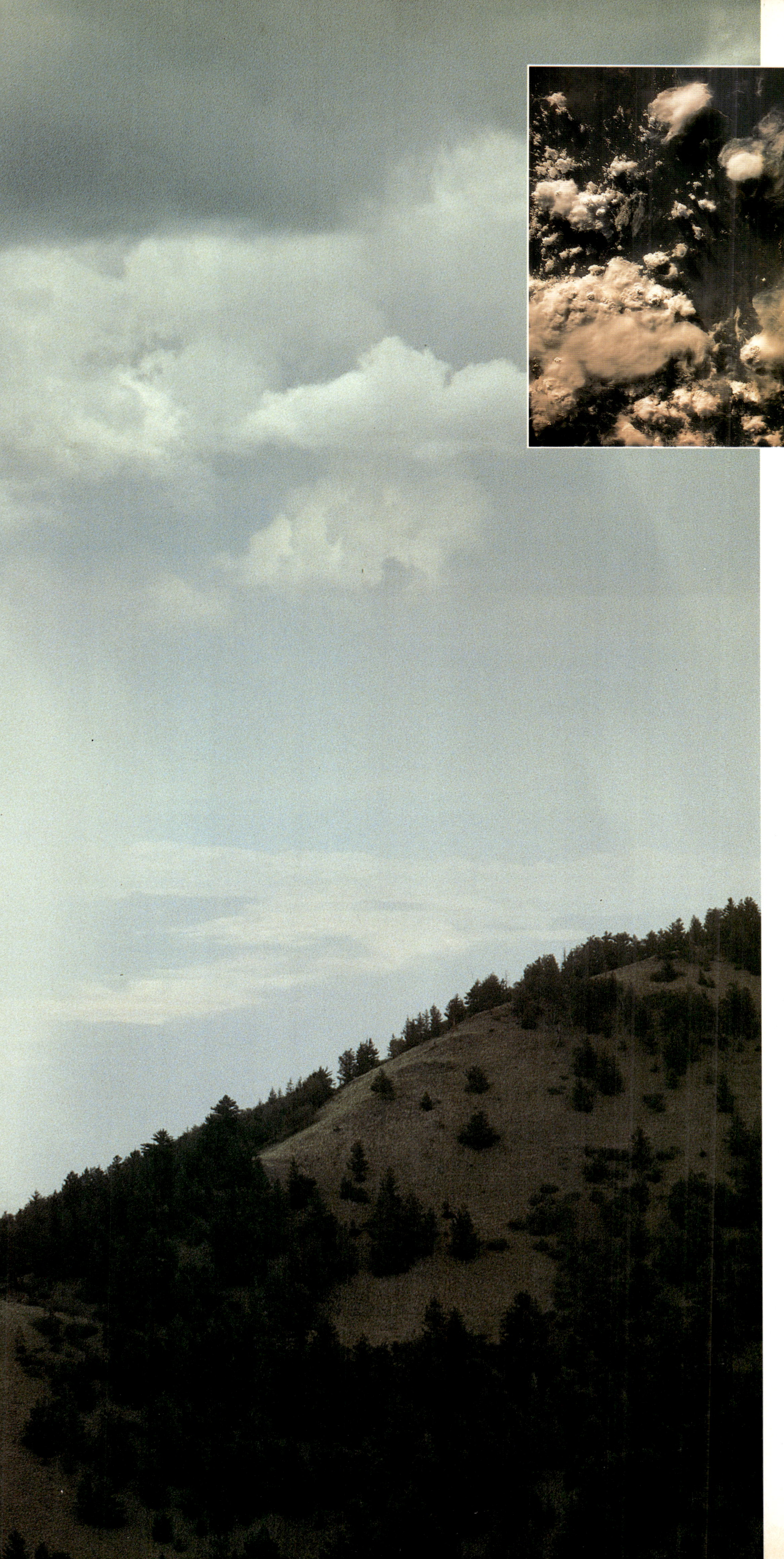

Above: Space observations have improved our understanding of large weather systems and our ability to track their movements. Thunderclouds (technically called cumulonimbus clouds) can extend from near the ground to 13 kilometers or more. They are heaped clouds, often resembling gigantic cauliflowers, usually topped by a flattened anvil created by winds at high altitudes. Violent updrafts and downdrafts and rolling motions are characteristic of these clouds. Thunderclouds, which may occur alone or in long lines along a front, are associated with some of the most violent weather we experience—torrential downpours, large hailstones, and gale-force winds. They can also spawn tornadoes and waterspouts.

Left: In the Magdalena Mountains of western New Mexico, weather and topographical conditions lead to the development of towering cumulus clouds and thunderstorms, particularly during July and August. The Langmuir Laboratory for Atmospheric Research, located in this region, is the site of a major research program focusing on how clouds form and the processes occurring within them. Researchers come from all over the world to study a wide variety of atmospheric and weather phenomena, including the electrical nature of thunderstorms, the fine structure and distribution of lightning, the relationship between lightning and rainfall, and the formation of hail. This is a good natural laboratory because thunderstorms here are often isolated, stationary, and relatively small; the clouds that form often go through an entire life cycle from birth to death without moving out of the area.

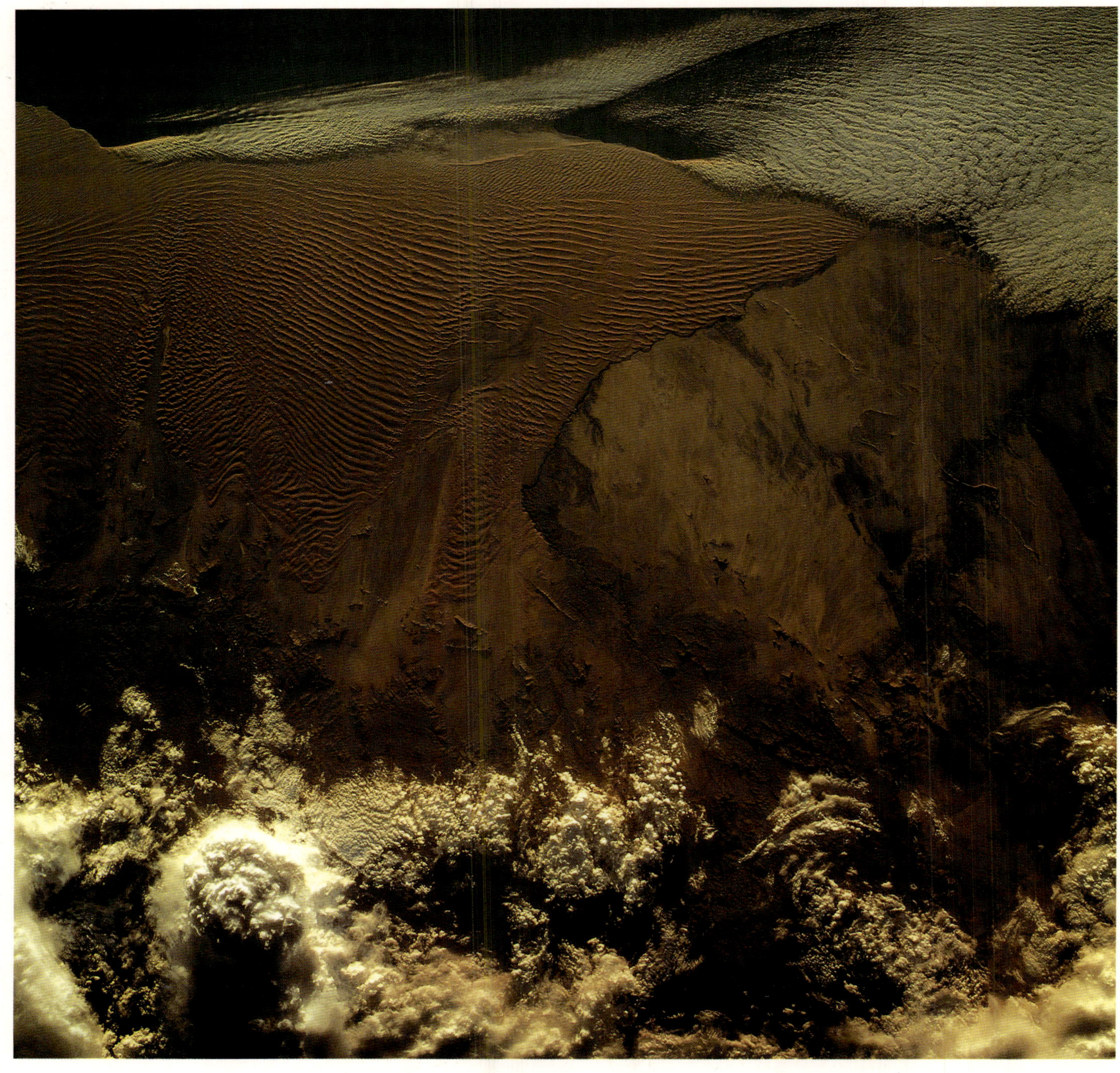

Above: The Namib, whose name means "an area where there is nothing," is a desert stretching along the Atlantic coast of Namibia in southwestern Africa. Brick-red dunes, some higher than 200 meters, run from north-northwest to south-southeast. A land of perennial drought, some regions do not get a drop of rain for years on end. However, fog from low-lying coastal clouds provides enough moisture to make life possible here. The Namib contains a unique collection of plants and animals.

Right: From space, the desert of Saudi Arabia appears to be peppered with hundreds of black circles. These are irrigation discs, 1 to 2 kilometers in diameter, which provide water needed for growing alfalfa. But wresting agricultural land from the desert comes with a high price tag—the Saudi farmers are forced to use fossil water, a nonrenewable resource. This is often referred to as "mining" the water because, in this arid region, the supply will not be replenished by rainfall. It will take only about fifty years to use up this water, but ten thousand years to replace it.

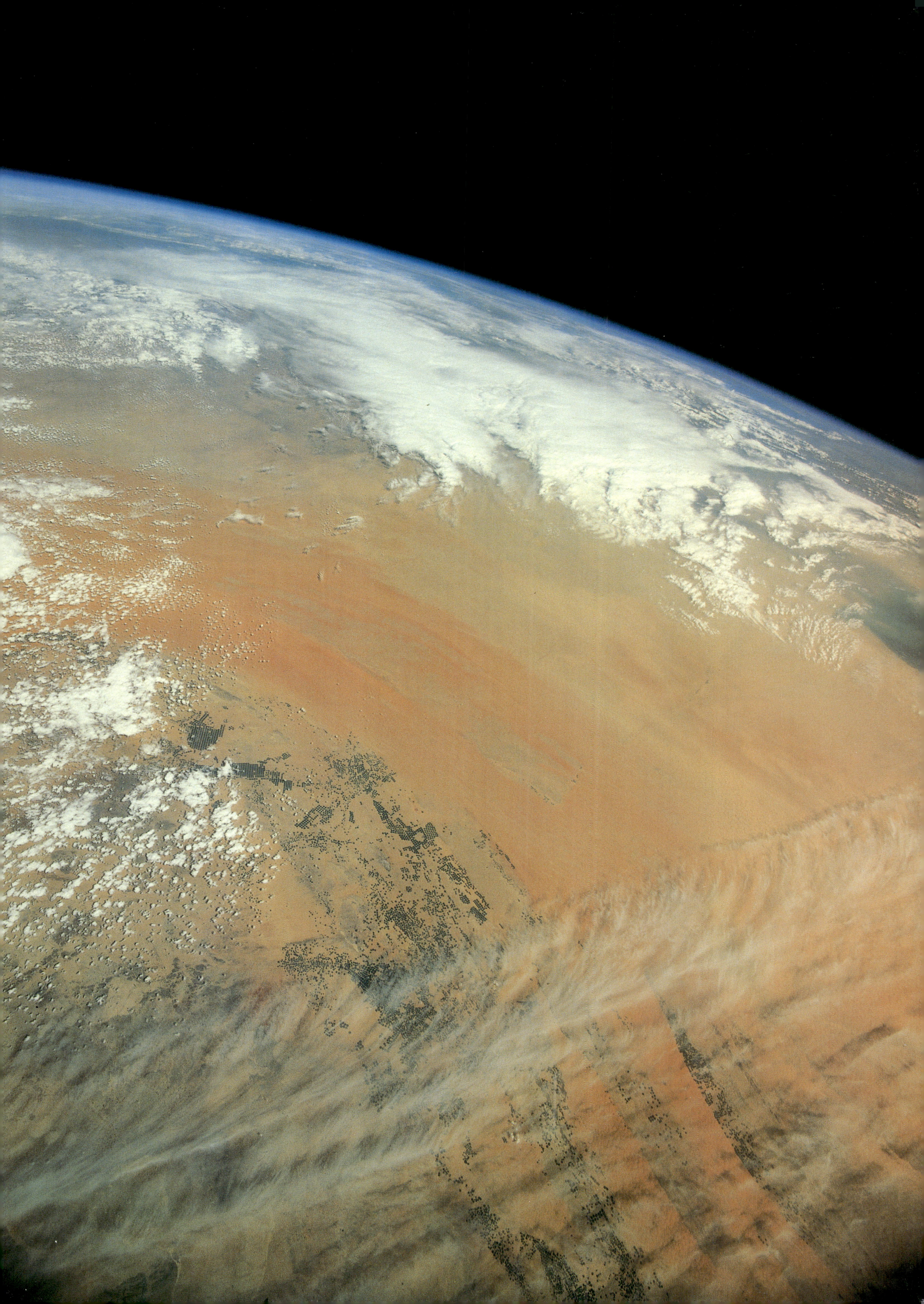

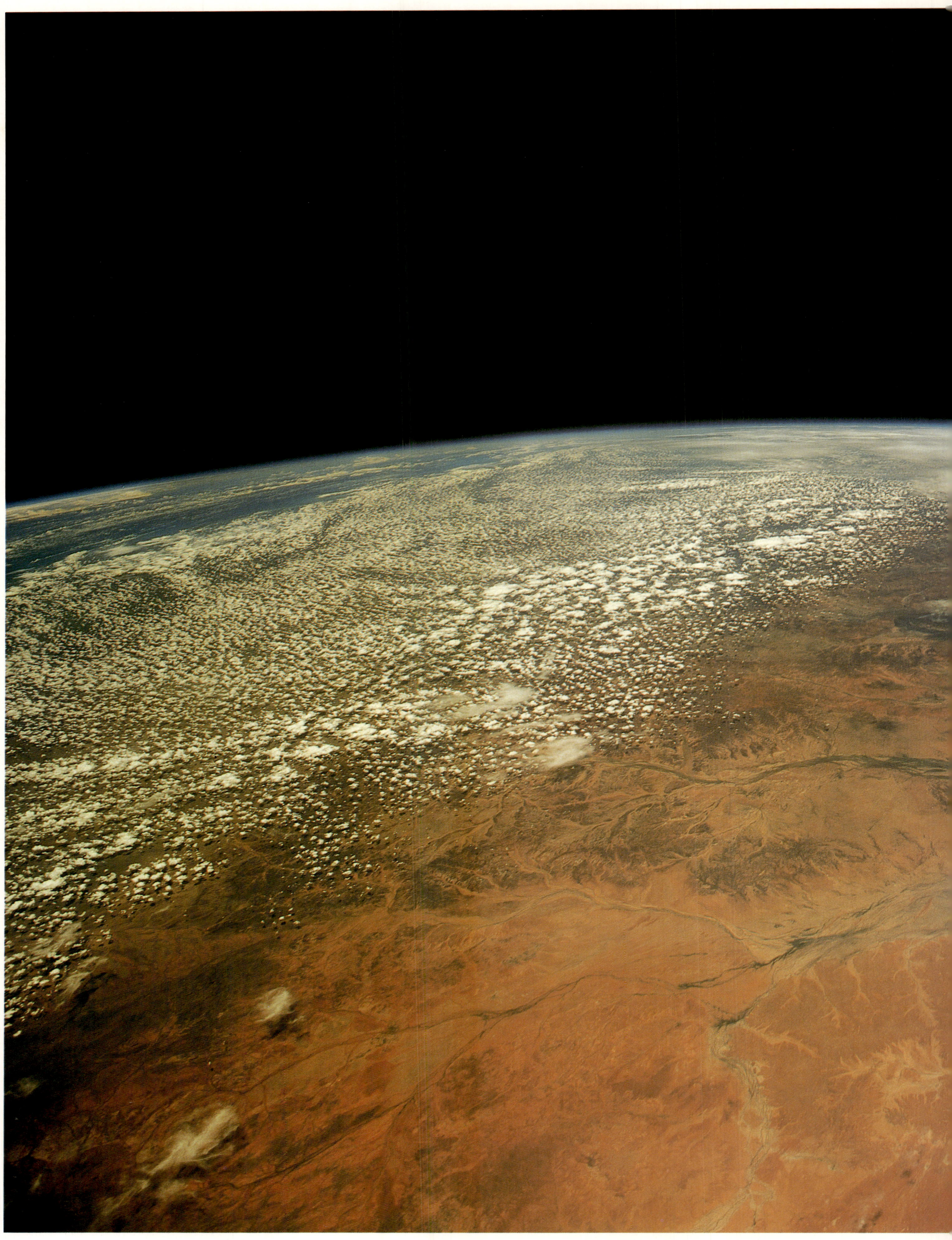

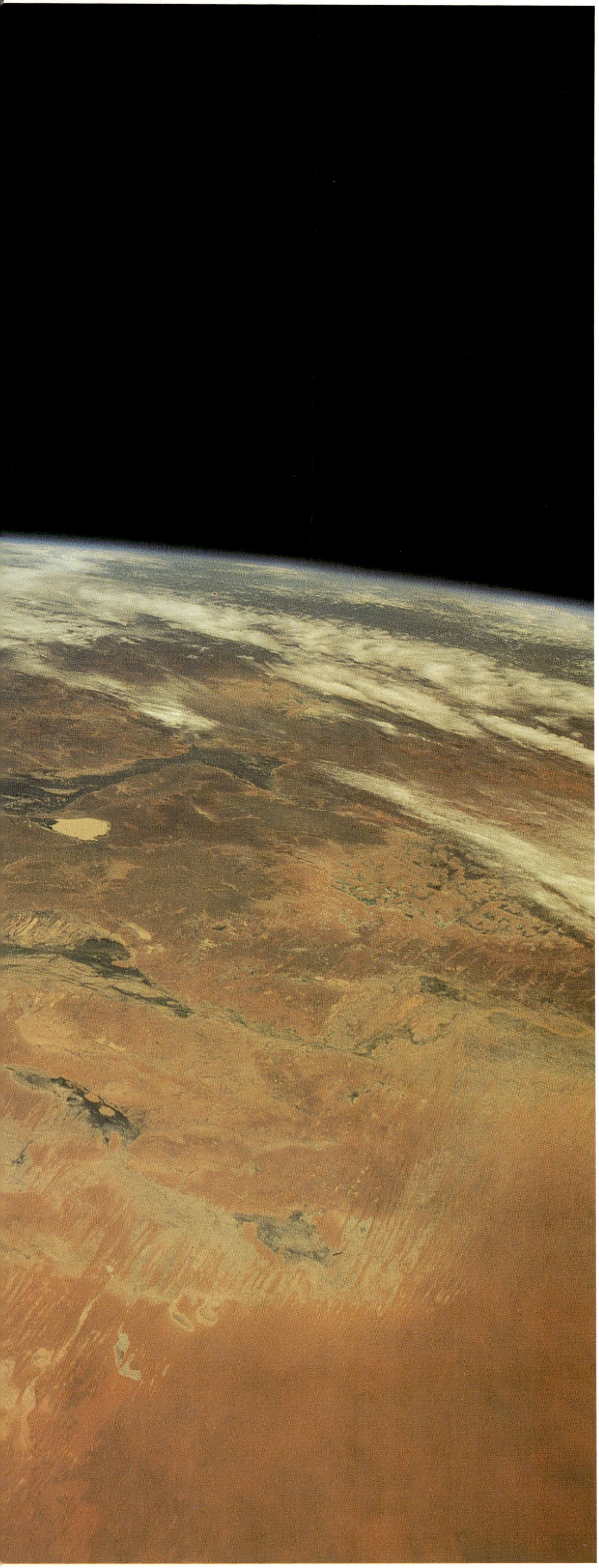

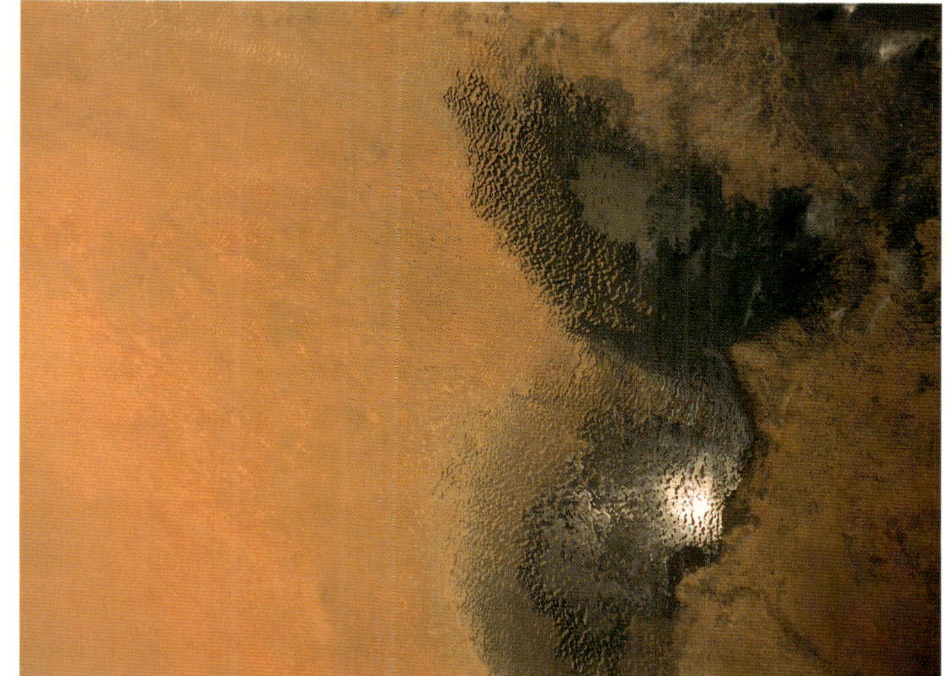

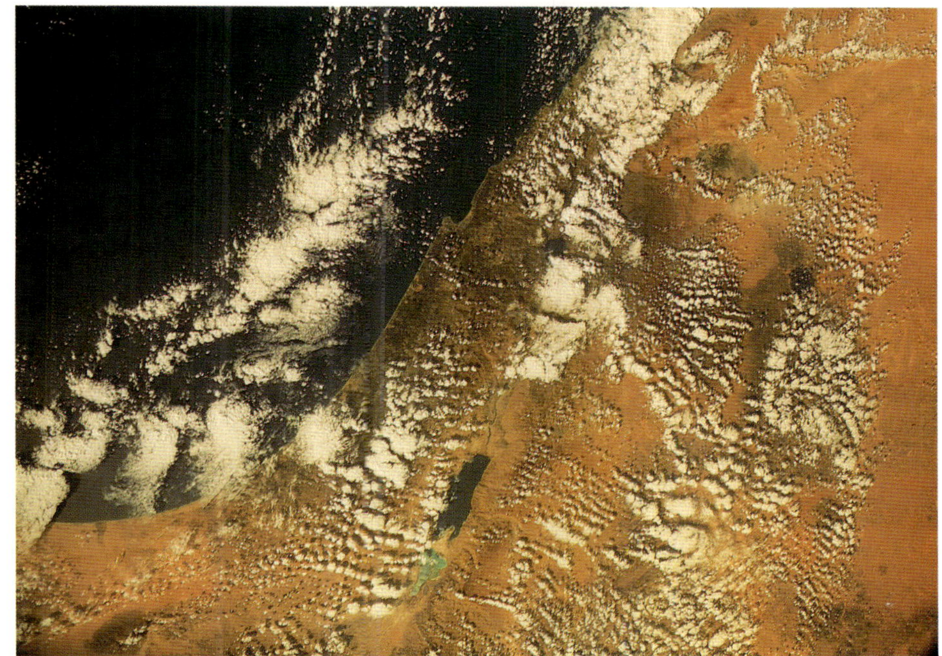

Left: This view looks over the northwestern and central regions of Australia, the flattest and driest continent on earth. Two-thirds of the country is desert or semidesert; most of the central region is dominated by vast, flat plains the color of dried blood, occasionally broken by striking, isolated rock formations. Australia's isolation from other continents accounts for its strange and distinctive animal species, many of which exist nowhere else on earth.

Top: Lake Chad, located in the Sahel in central Africa, is a favorite target of astronauts' cameras. Its name, which means "great body of water," was more apt thousands of years ago when the lake extended over as much as 330 thousand square kilometers. For most of historical time, it has covered less than 10 percent of that area, and it has receded greatly in recent years, particularly during the severe drought between 1968 and 1974. The ghost of Lake Chad's past can be seen etched into the sand crowding its shores. Ancient dunes, once submerged, are now exposed.

Above: The Jordan River rift valley extends through the Dead Sea (center bottom) along the length of Israel's eastern boundary. The river connects the Dead Sea with the Sea of Galilee, partially visible beneath cloud cover to the north. The Dead Sea, 401 meters below sea level, is the lowest point on the earth's surface; its southern basin appears lighter blue because of its shallow depths. The southern part of Israel is dominated by the Negev Desert.

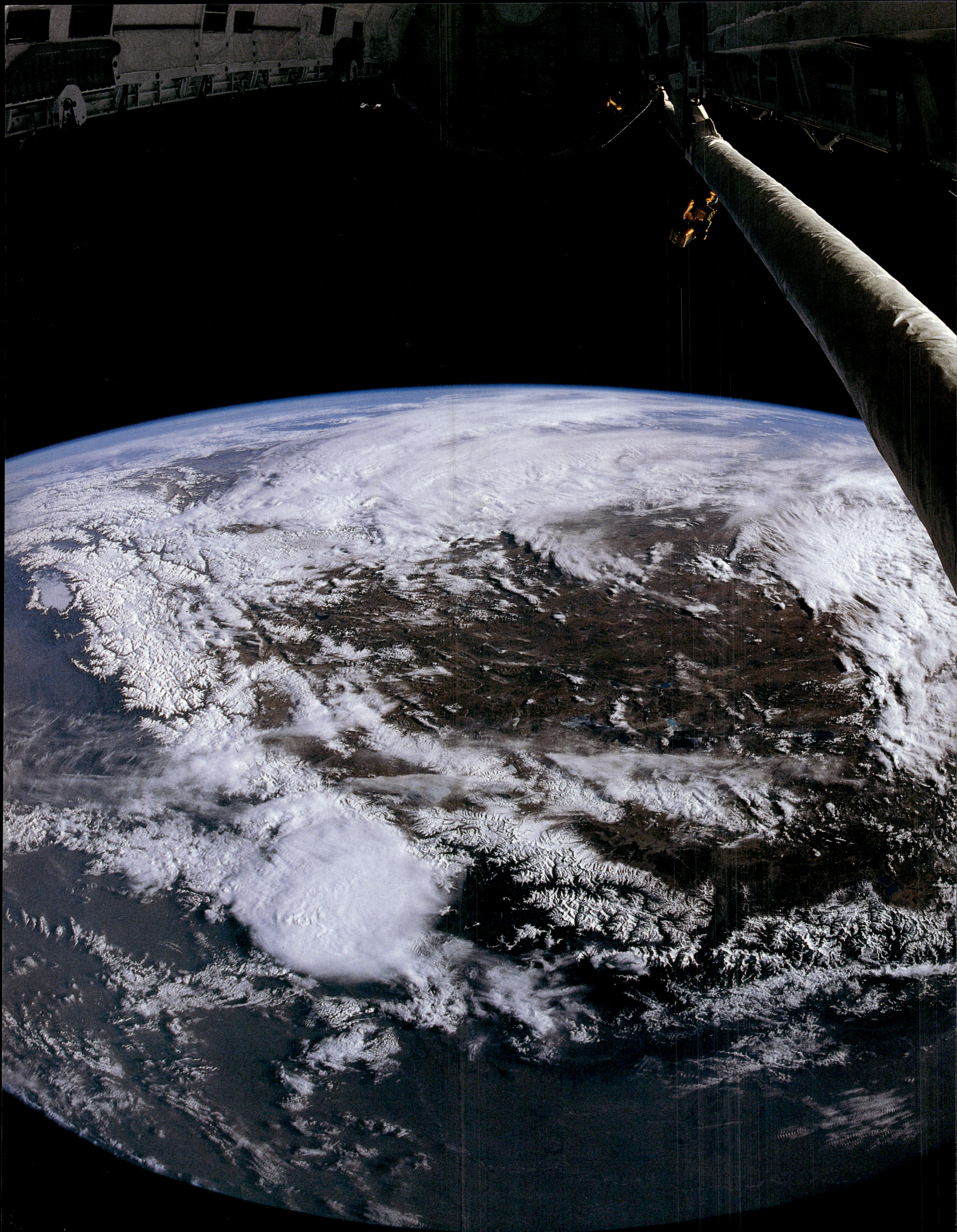

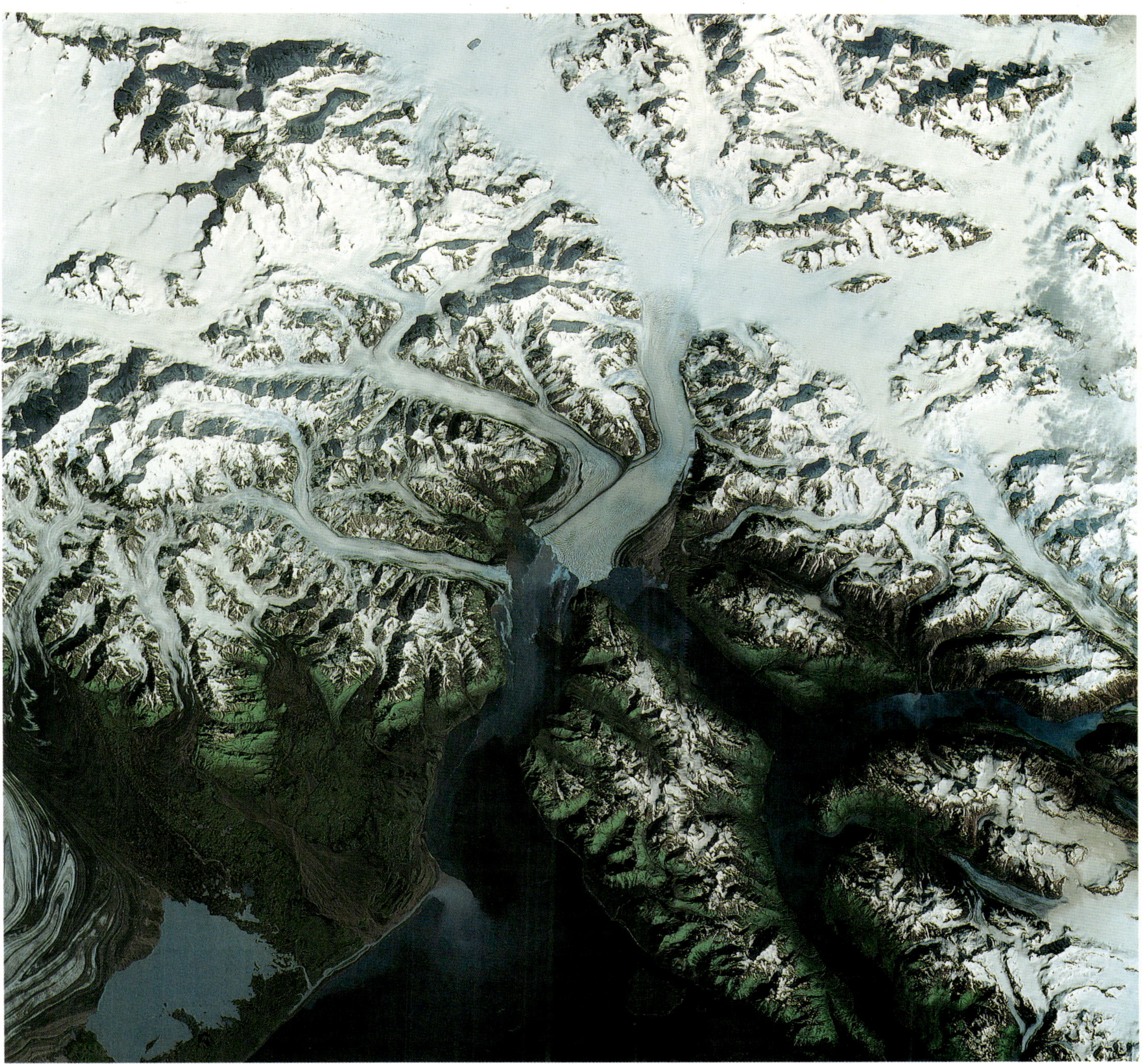

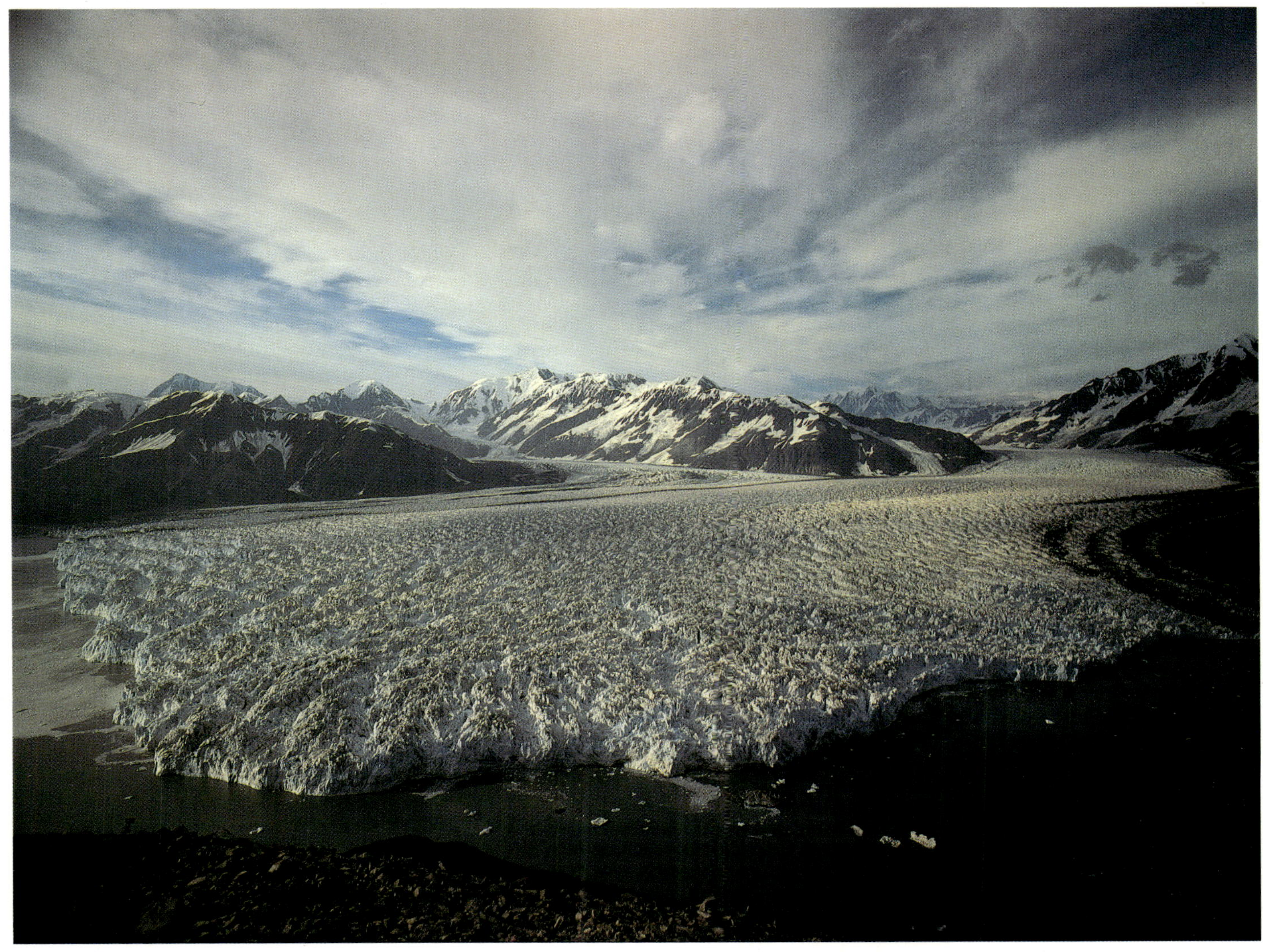

Left: The Himalaya Mountains stretch uninterruptedly for about 2,500 kilometers along the northern border of India, providing an almost impassable barrier between the Indian subcontinent and China and Tibet. The mountains have been thrust up by the slow crushing together of the crustal plate that carries India northward and the larger plate carrying Asia. The Himalaya range boasts more than thirty peaks above 7,300 meters; Mount Everest, a perpetual challenge to mountaineers, stands at 8,848 meters—the tallest mountain in the world, measured from sea level. The mountains have a significant effect on the climate of the region; they block cold continental air from the north and force monsoon winds from the south to dump their moisture as heavy rain and snow on the Indian side, leaving Tibet extremely arid.

Above right: Hubbard Glacier in Alaska is the longest valley glacier in North America, stretching more than 150 kilometers from the Yukon Territory to Yakutat Bay, which feeds into the Gulf of Alaska (bottom). In this Landsat satellite image, Hubbard can be seen encroaching almost halfway down the bay. Hubbard has advanced and retreated over the centuries and at times has blocked the entrance to Russell Fjord, which branches to the right. Two smaller glaciers can be seen to the left of Hubbard; the upper, Valerie, joins Hubbard while the lower, Turner, spills into Yakutat Bay. Ice sheets are important in regulating the earth's climate because of the extent to which they reflect the sun's energy back into space.

Right: The leading edge of Hubbard Glacier and that of its smaller companion, Valerie Glacier, can be seen in this photo. Inside these walls of ice is a climatic record going back thousands of years. Glaciers such as these are small remnants of the great ice sheets that covered much of North America more than ten thousand years ago during the last ice age. In the past million years, there has been a succession of ice ages interspersed with relatively brief warm periods called interglacials like the one earth is experiencing now. Glaciers advanced and retreated with the changing climate, altering the earth's surface.

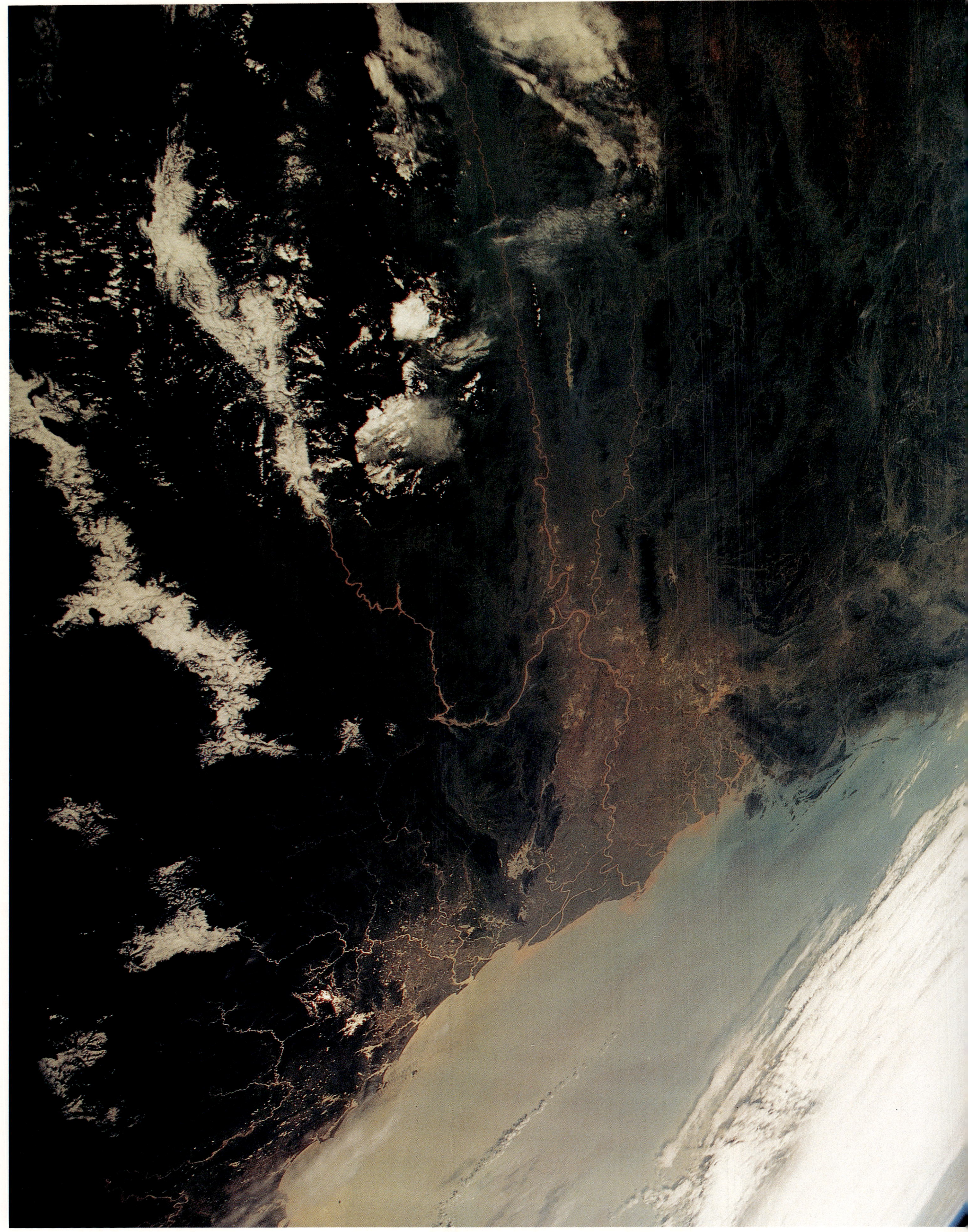

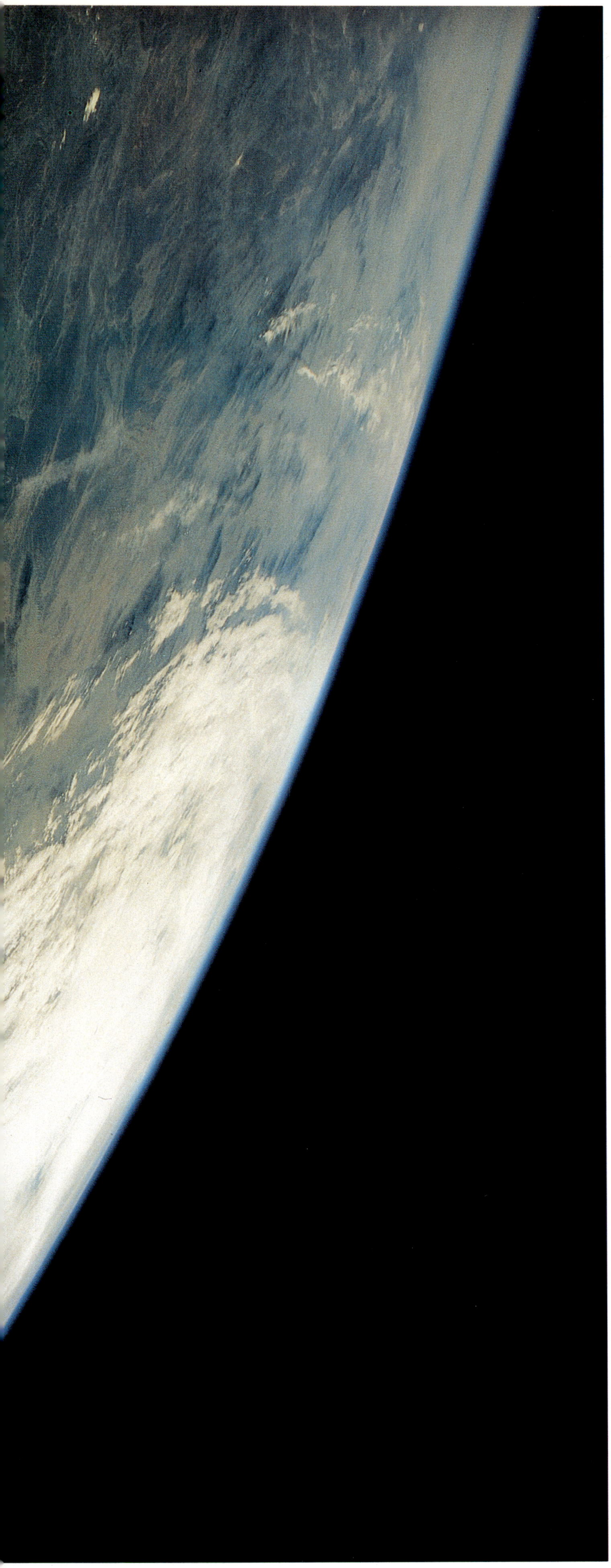

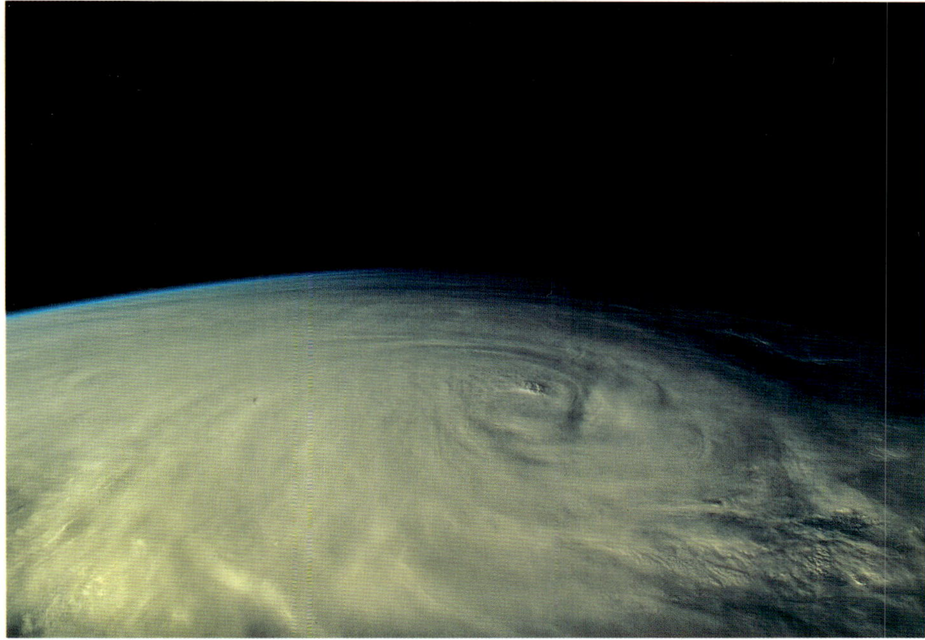

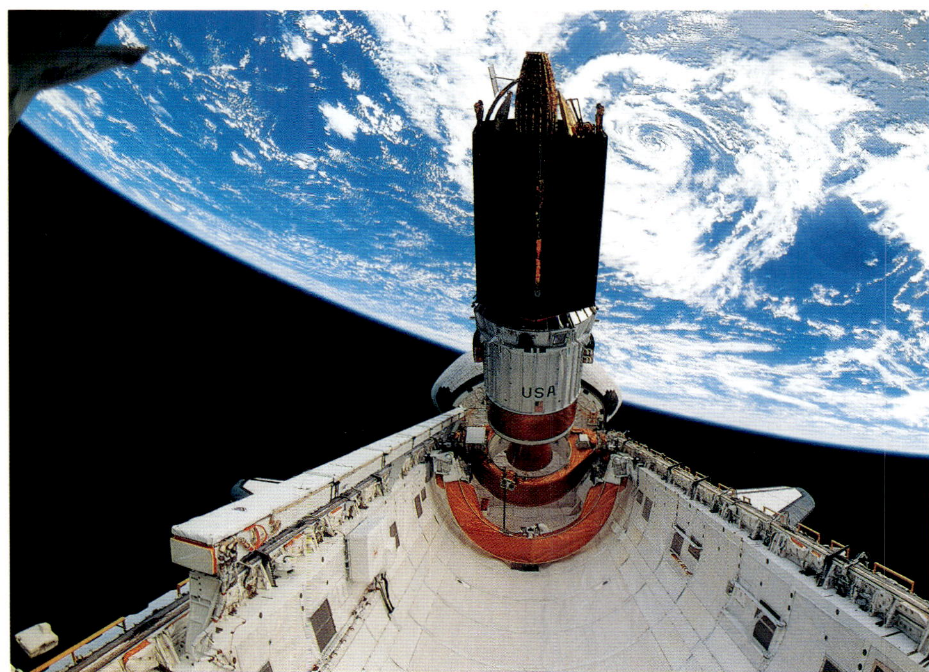

Left: Southeast Asia has the highest frequency of violent cyclonic storms in the world. Here, the coast of Vietnam can be seen with a typhoon hovering offshore. In the center is the mouth and the delta of the Hong, or Red, River, where Hanoi and Haiphong Harbor are located. In 1881, when a typhoon hit Haiphong, 300 thousand people died, mostly by drowning.

Top: During a flight in January, 1990, shuttle astronauts photographed Typhoon Sam near Australia. Cyclonic storms are extreme low-pressure systems. In the southern hemisphere, they revolve in a counterclockwise direction, opposite to that of similar storms in the northern hemisphere. Space observations have greatly improved the forecasting of these storms, which can take several days to reach land. Knowing more precisely when and where they will come ashore helps in the planning of evacuations and in taking measures to reduce property damage. In North America and the Caribbean, these cyclonic storms are called hurricanes.

Above: The Tracking and Data Relay Satellite (TDRS) is launched from the cargo bay of the space shuttle *Discovery* in March, 1989. Eventually positioned in geosynchronous orbit 35,800 kilometers above the equator over the Atlantic Ocean, TDRS is one of a family of advanced communications satellites designed to provide improved information transfer among other satellites and the ground and also to enhance communications and relay data during shuttle missions. Immediately behind the satellite, a whirlpool-like cloud system can be seen. These cyclonic structures can sometimes grow huge enough to blanket half a continent or more; warm and cold fronts along their spiral arms are the sites of unsettled and violent weather. The vantage point of space has enabled scientists to observe such large weather systems as a whole, to track their movements, and to watch how they change over time.

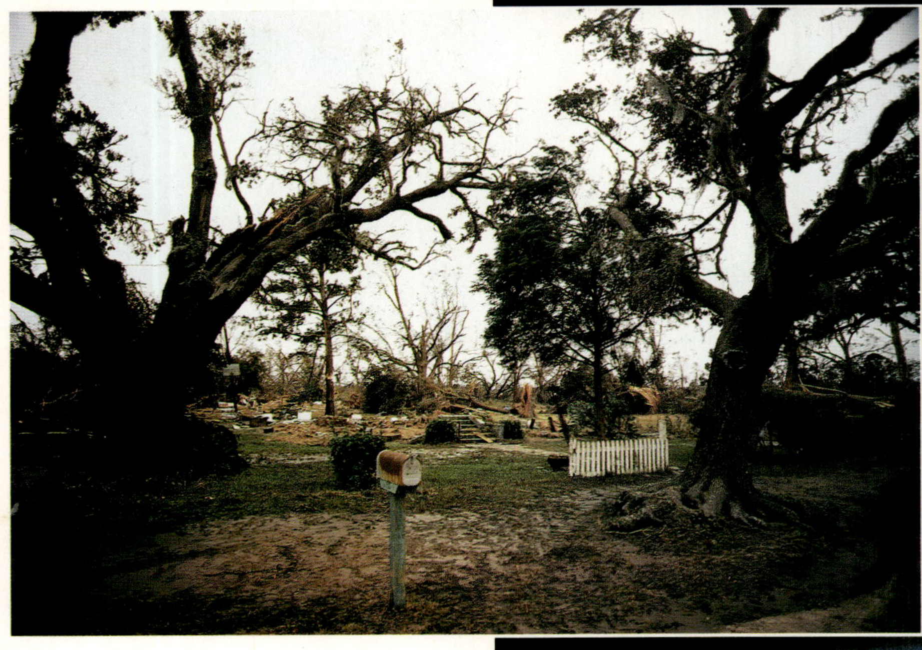

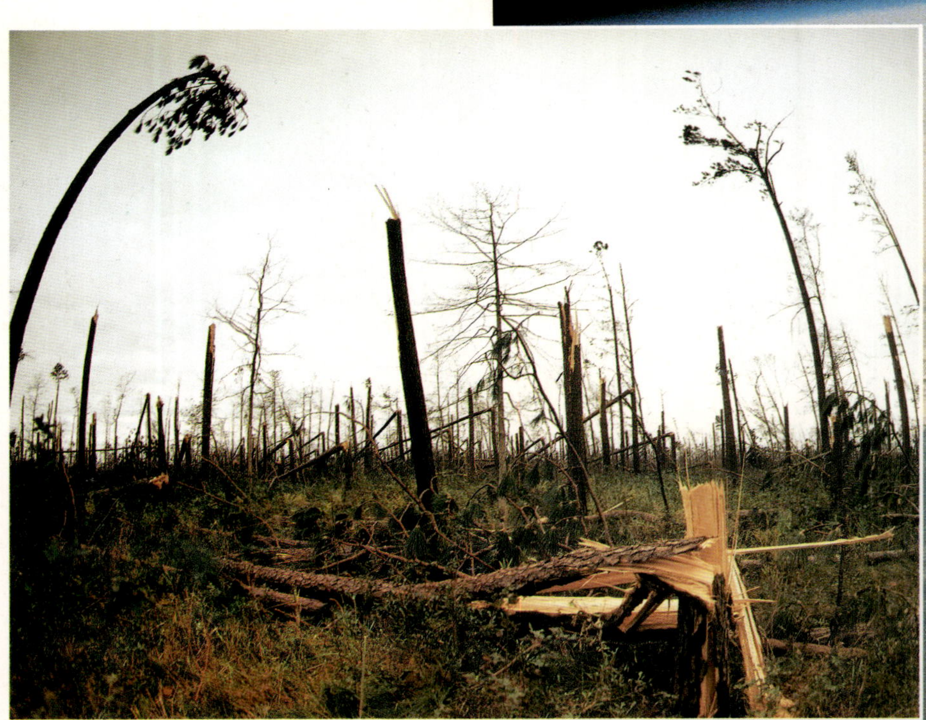

Top: By definition, hurricane winds exceed 117 kilometers per hour. Much of the devastation they cause results from the winds, but hurricanes also create storm surges—a raising of sea levels caused by the high winds and low pressure over the ocean surface. The sea levels can rise several meters, enough to inundate low-lying coastal regions and cause severe flooding.

Above: Humans are not the only ones who suffer from hurricanes. When Hugo hit South Carolina packing winds nearly 240 kilometers per hour, it devastated the Francis Marion National Forest in South Carolina and, along with it, the forest's population of red-cockaded woodpeckers, an endangered species. When the storm was over, most of the woodpeckers were dead or missing, and the others were further endangered by the loss of their homes. Red-cockaded woodpeckers spend nearly five years pecking holes in trees to create their nests, which are handed down from generation to generation. Ironically, the holes weakened trees and made them vulnerable to Hugo's fury.

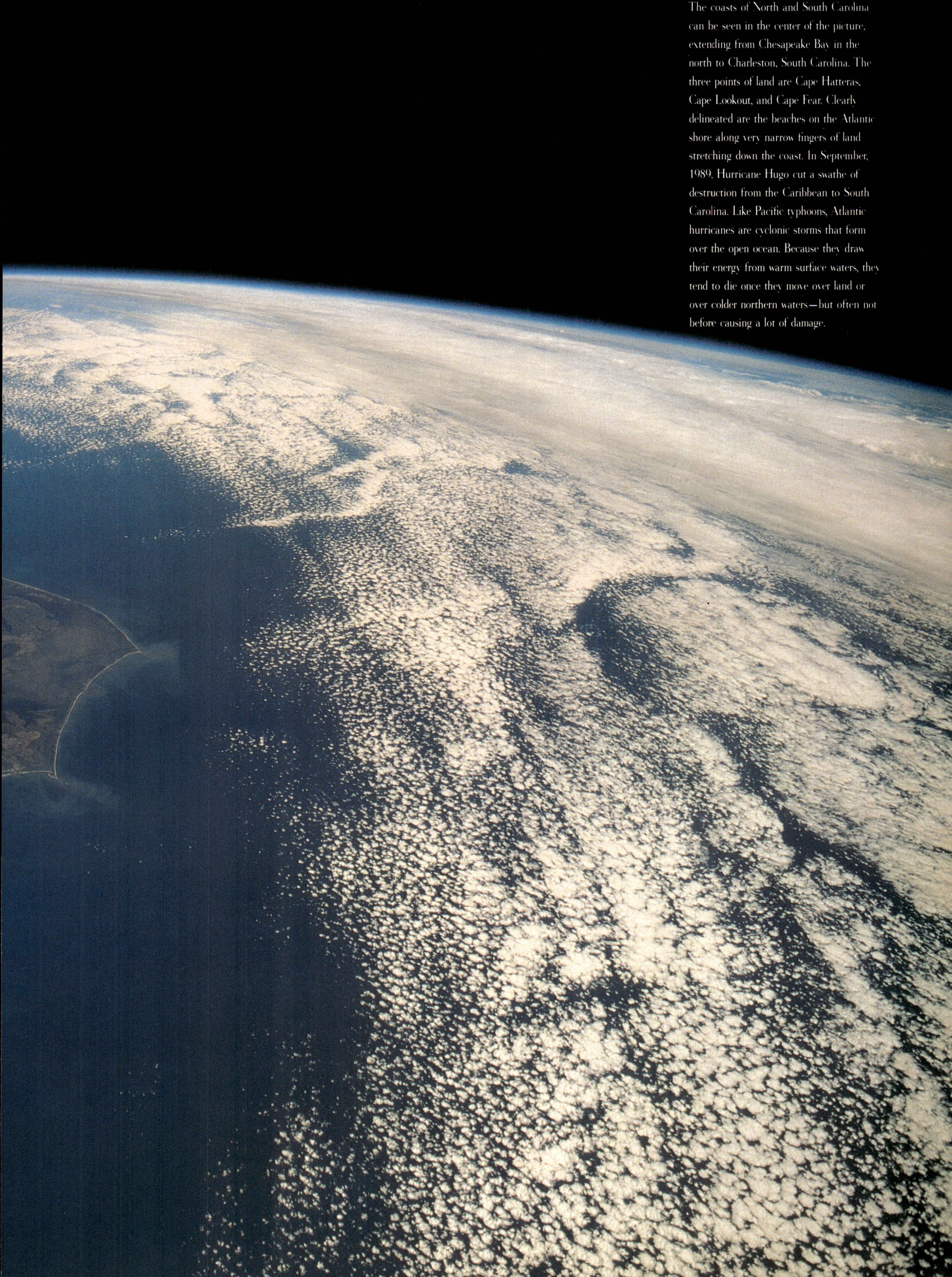

The coasts of North and South Carolina
can be seen in the center of the picture,
extending from Chesapeake Bay in the
north to Charleston, South Carolina. The
three points of land are Cape Hatteras,
Cape Lookout, and Cape Fear. Clearly
delineated are the beaches on the Atlantic
shore along very narrow fingers of land
stretching down the coast. In September,
1989, Hurricane Hugo cut a swathe of
destruction from the Caribbean to South
Carolina. Like Pacific typhoons, Atlantic
hurricanes are cyclonic storms that form
over the open ocean. Because they draw
their energy from warm surface waters, they
tend to die once they move over land or
over colder northern waters—but often not
before causing a lot of damage.

The earth has had a hard and violent past. Its surface is weathered and wrinkled, scarred with impact craters, with mountain chains uplifted by colliding plates, and with huge rifts torn open by earthquakes. Even today, after 4.5 billion years, it remains a dynamic place, its face constantly rearranged by the pounding of wind and water and by ongoing geological upheaval in the form of earthquakes and volcanoes.

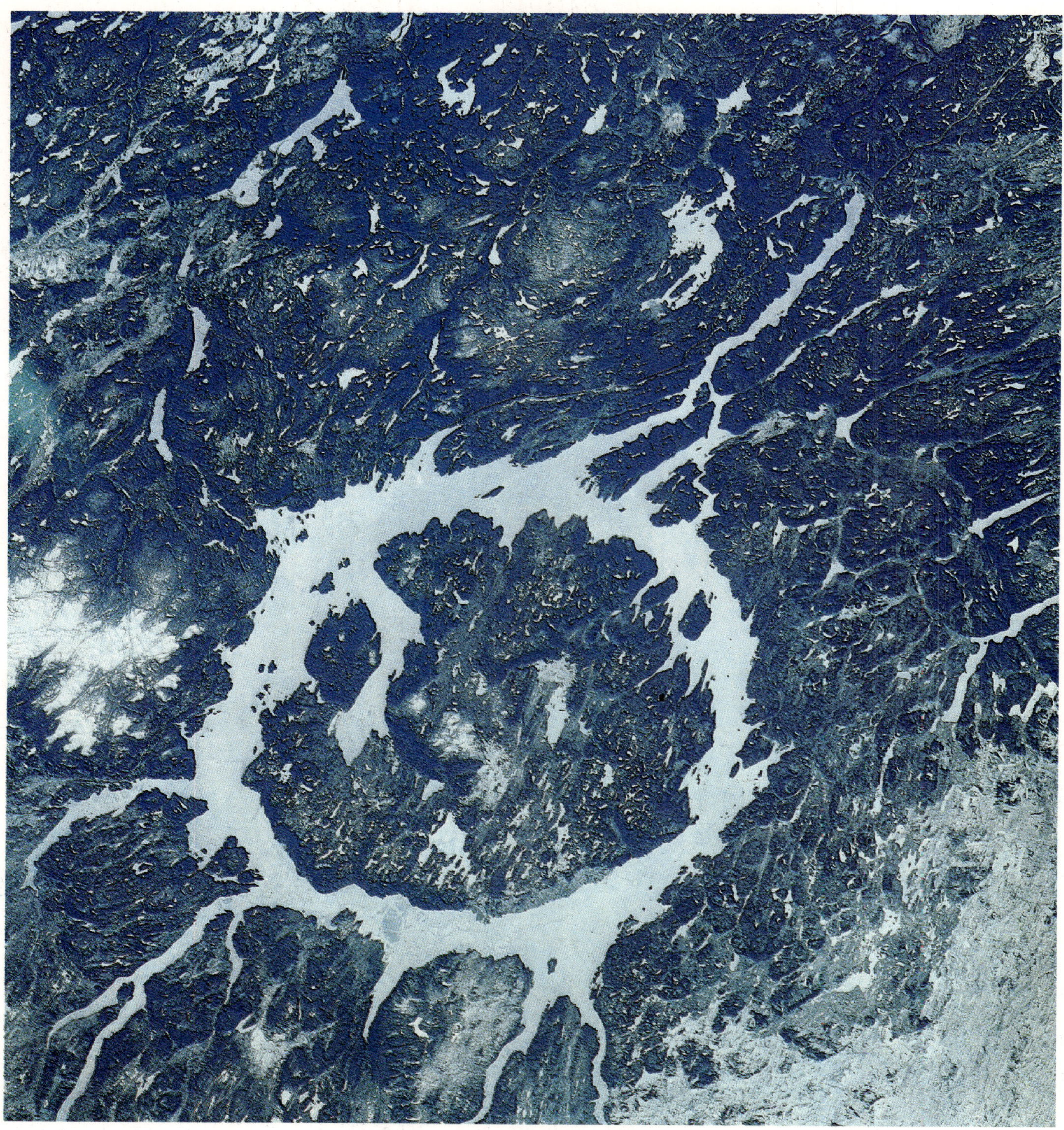

Most meteorites that encounter earth burn up in the atmosphere, but those that survive the fiery entry impact the surface with great violence. The crater at right near Flagstaff, Arizona was created by an impact that excavated more than 300 million tons of rock and destroyed surrounding plant and animal life. Even after thirty thousand years, the crater is still nearly 180 meters deep. In the image above, Manicouagan Crater in east-central Quebec, 60 kilometers across, is sharply outlined by ice-covered lakes. Its deeper structure has been exposed by the scouring of glaciers. In its early history, earth's surface was covered with craters but most have since been erased by erosion and internal geological forces. Scientists have suggested that climatic changes caused by a huge impact may have caused the massive extinction of species, including all the dinosaurs, that occurred about sixty-five million years ago.

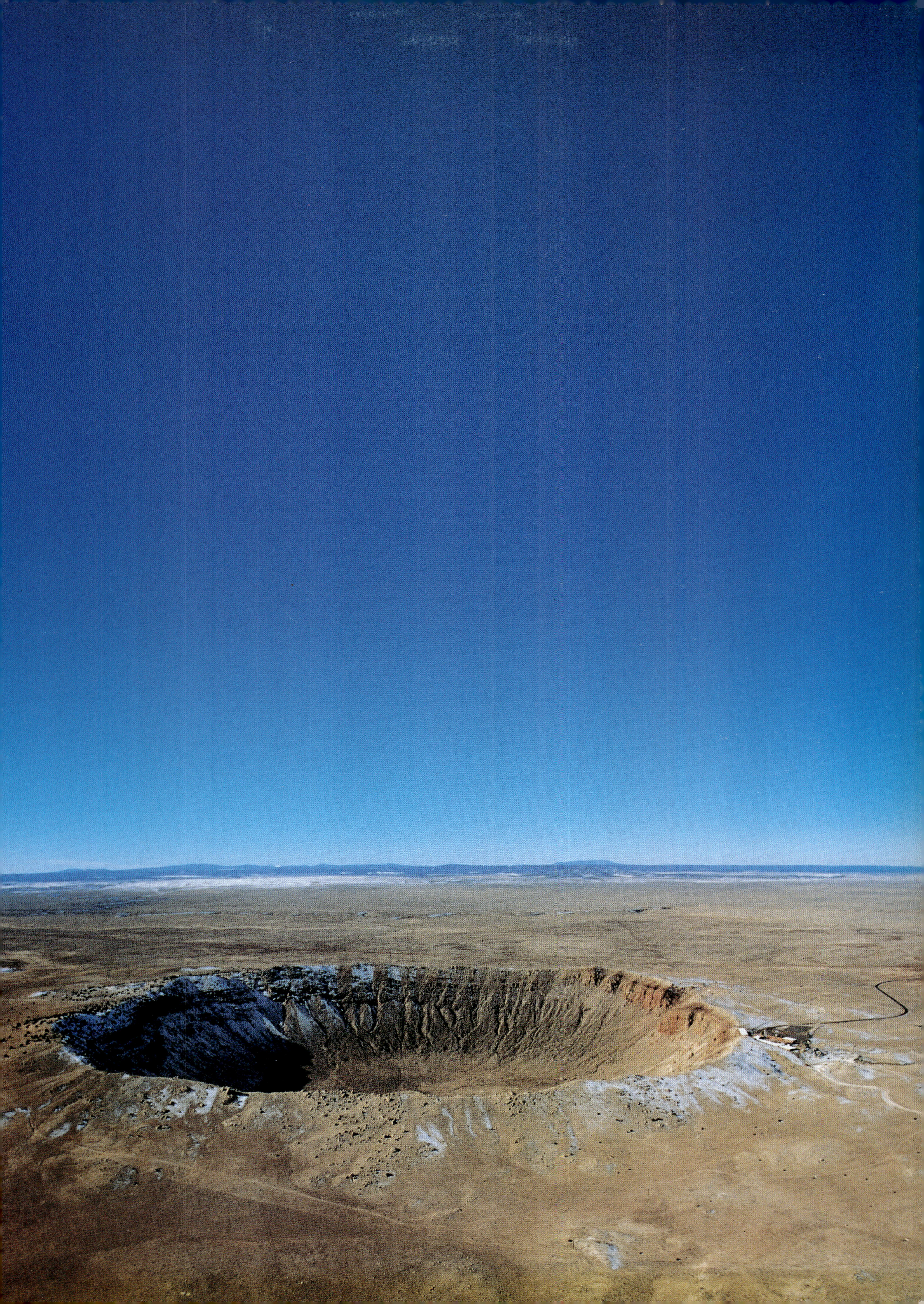

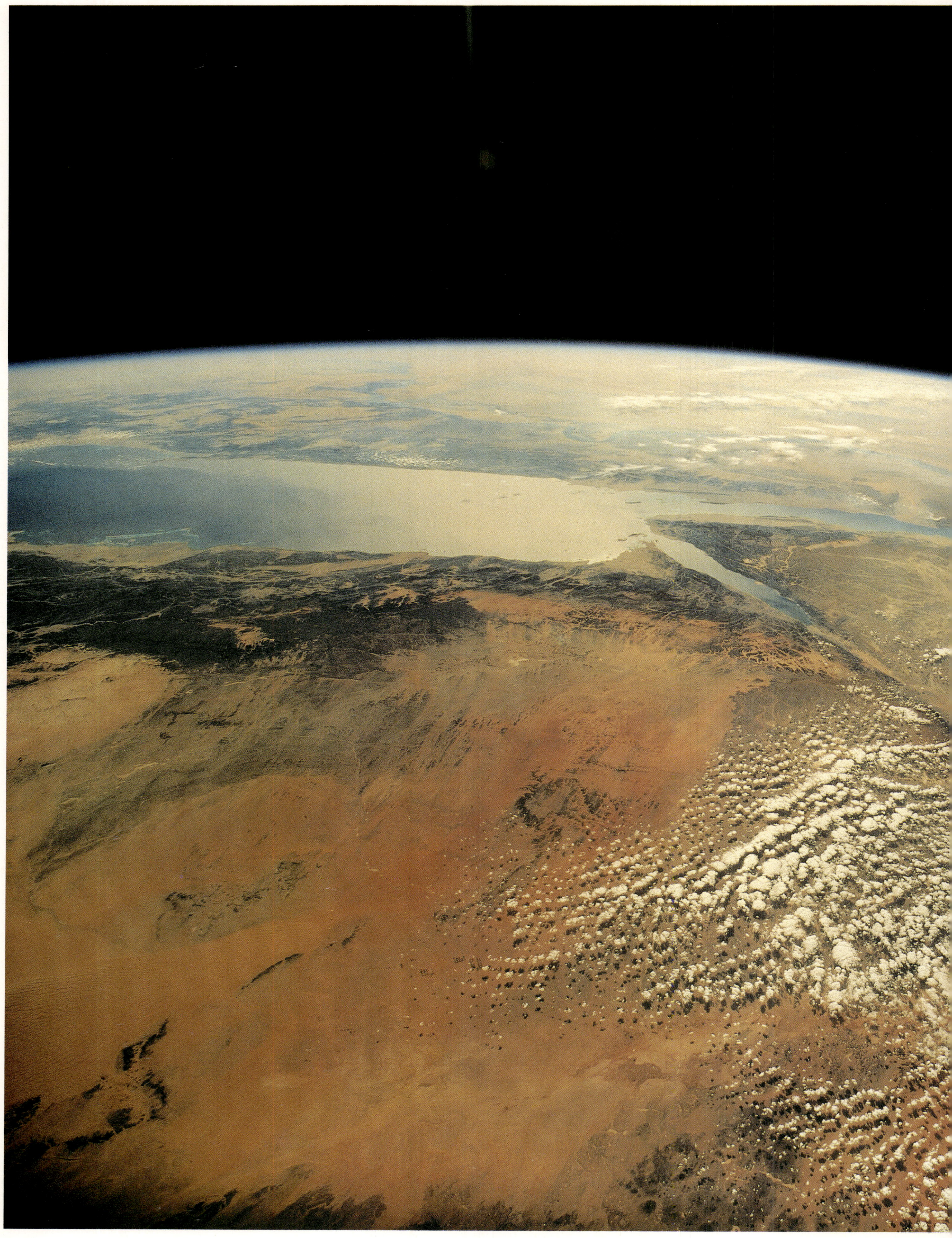

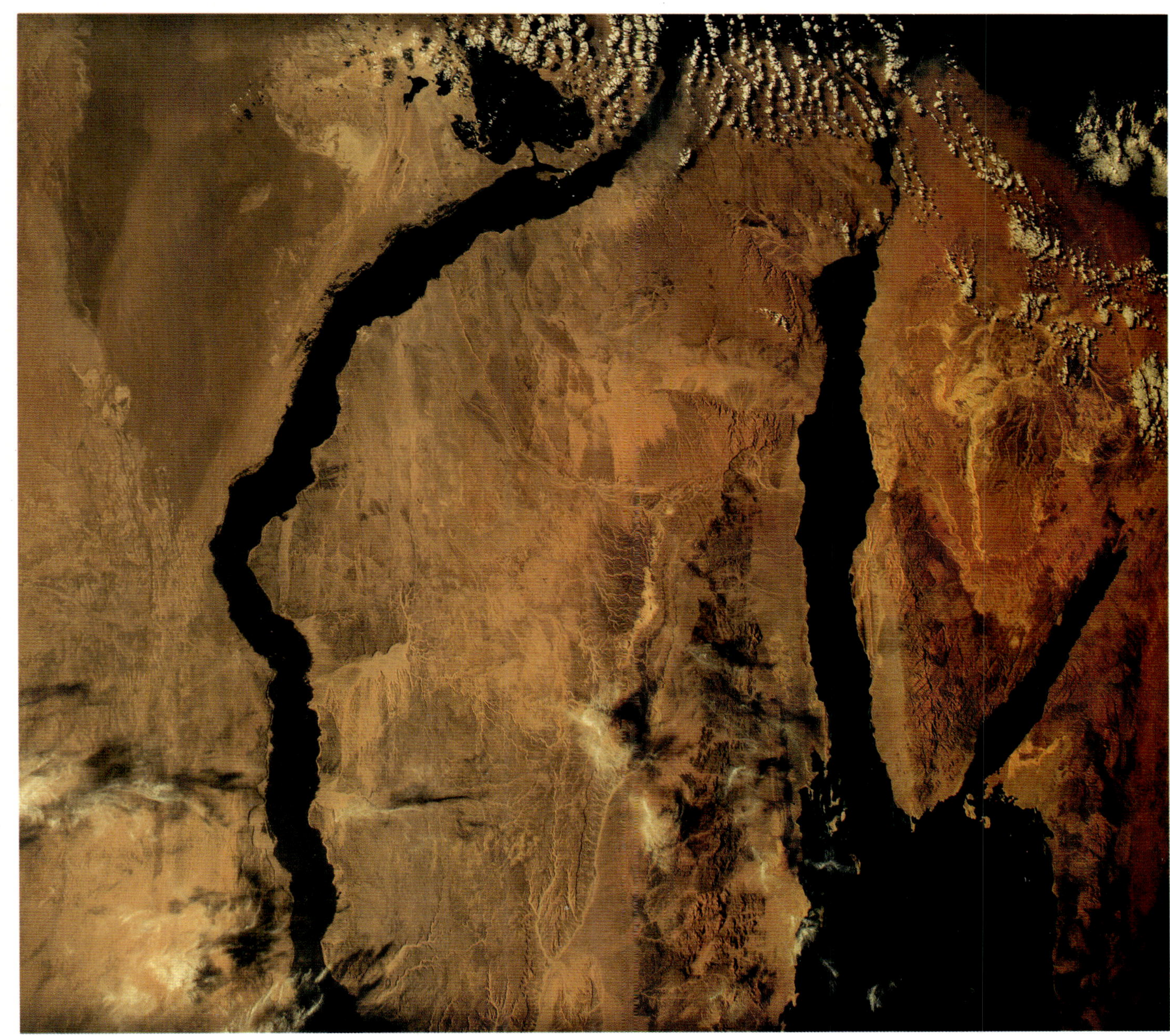

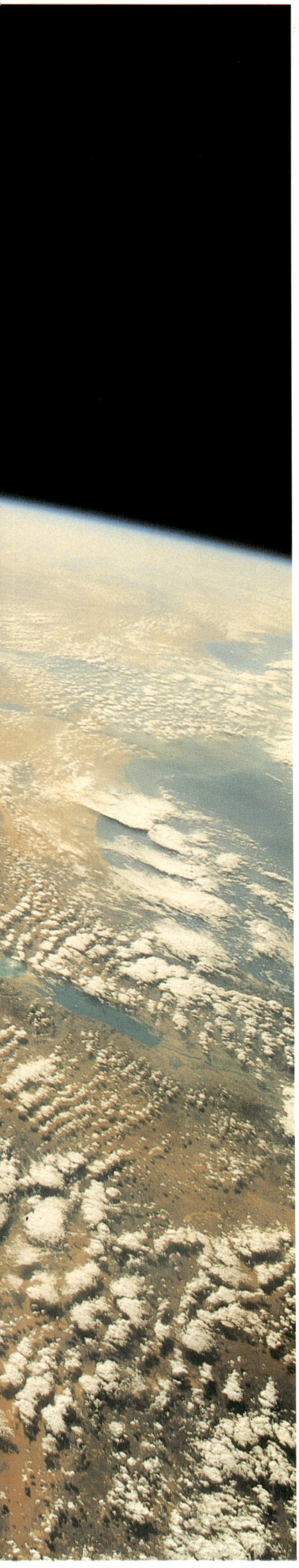

Left: This picture looks west over Saudi Arabia and the Red Sea into Egypt. On the right, the V-shaped point of the Sinai Peninsula separates the Gulf of Aqaba (foreground) from the Gulf of Suez. North of the Gulf of Aqaba is the Jordan River rift valley, which opened when the Sinai Peninsula wrenched away from Saudi Arabia. One of the few crustal rifts (sections where the earth's crust has separated) exposed on land, it provides dramatic evidence of a major geological discovery—the fact that the earth's crust is broken into huge moving plates.

Above: Most of the landmass of Egypt is covered by the Sahara Desert. The Nile River (left), the longest river in the world, supports most of the country's population and agriculture. Dark areas along the river's banks and the distinctive leaf-shaped region near the top of the picture indicate where irrigation has permitted agricultural productivity. To the right is the Sinai Peninsula, flanked by the Gulfs of Suez and Aqaba. The Suez Canal can be seen at top right.

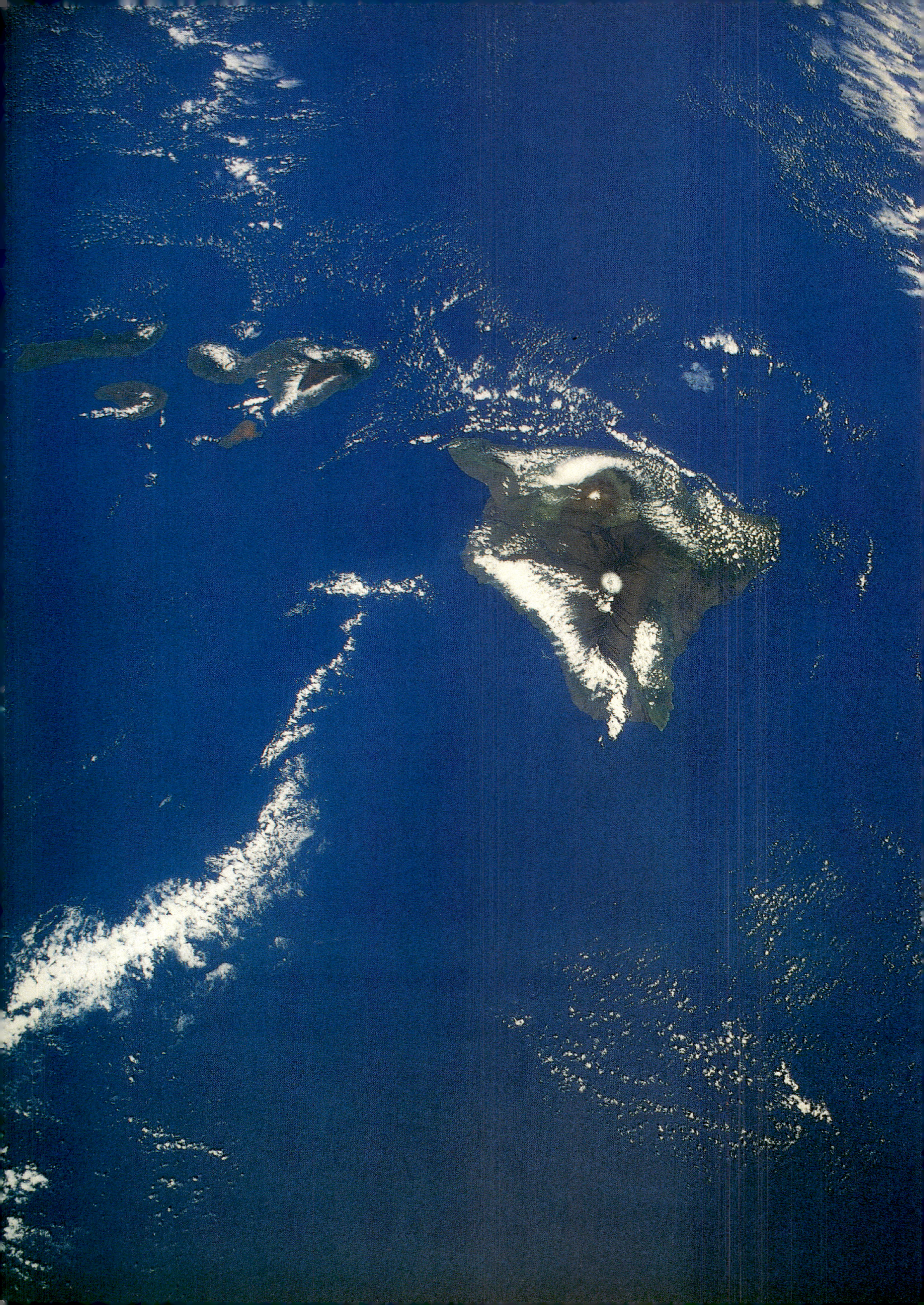

Left: The Big Island of Hawaii, the youngest of the Hawaiian islands, was formed by the outpouring of lava from five volcanoes, two of which—Mauna Loa and Kilauea—are still actively adding land area to the island today. Dark-colored lava flows can be seen spreading down the flanks of Mauna Loa, whose snow-capped peak is near the center of the island. Mauna Kea, the extinct volcano further north, is the tallest mountain on earth, when measured from its base on the seafloor to its peak (9,750 meters). Telescopes located on its summit take advantage of the altitude to escape the obscuring effects of the earth's atmosphere. Northwest of the Big Island are Maui, Kahoolawe, Lanai, and Molokai.

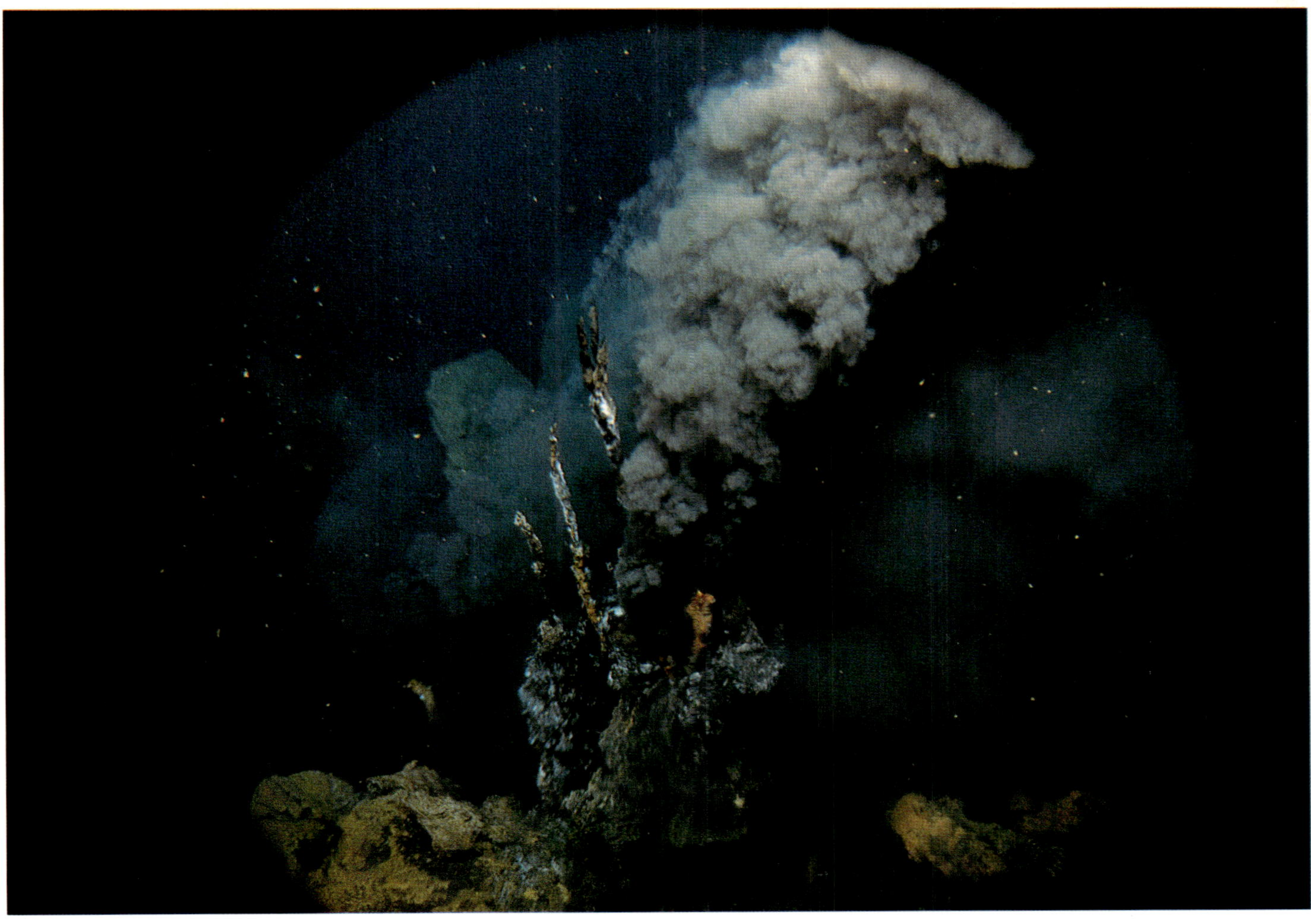

Above right: Superheated water escapes from vents on the ocean floor, carrying with it a soup of chemicals that rains down from plumes resembling black smoke. These chimneys are in fact called black smokers. Amazingly, a wide variety of animals— clams, snails, crabs, worms, and small shrimp-like creatures—are able to live in this dark and acidic environment with the help of bacteria that provide the basis of the food chain around the vents; these bacteria produce food from the chemicals spewed out by the smokers.

Right: The heat generated by the molten rock deep beneath the earth's surface creates oases of life in unexpected places in the ocean depths. Here, the French submersible *Nautile* prepares to dive to the floor of the Pacific Ocean to explore volcanic thermal vents. First discovered in 1977, these vents stunned scientists because they support thriving communities of strange life-forms, such as blind crabs and giant tube worms, in a habitat characterized by crushing pressures and a complete absence of sunlight for photosynthesis. Since then, scientists have discovered many other deep-sea vent communities.

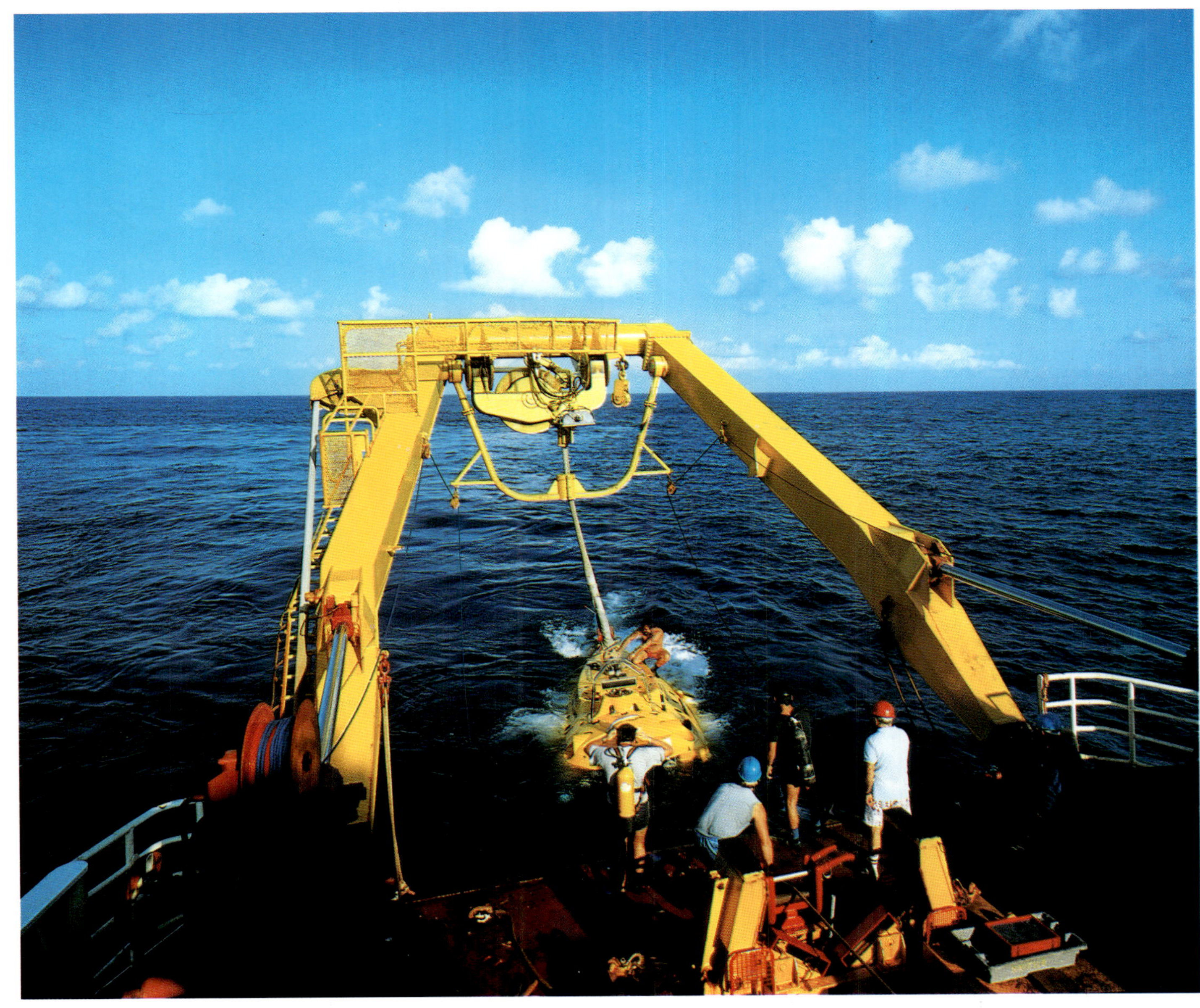

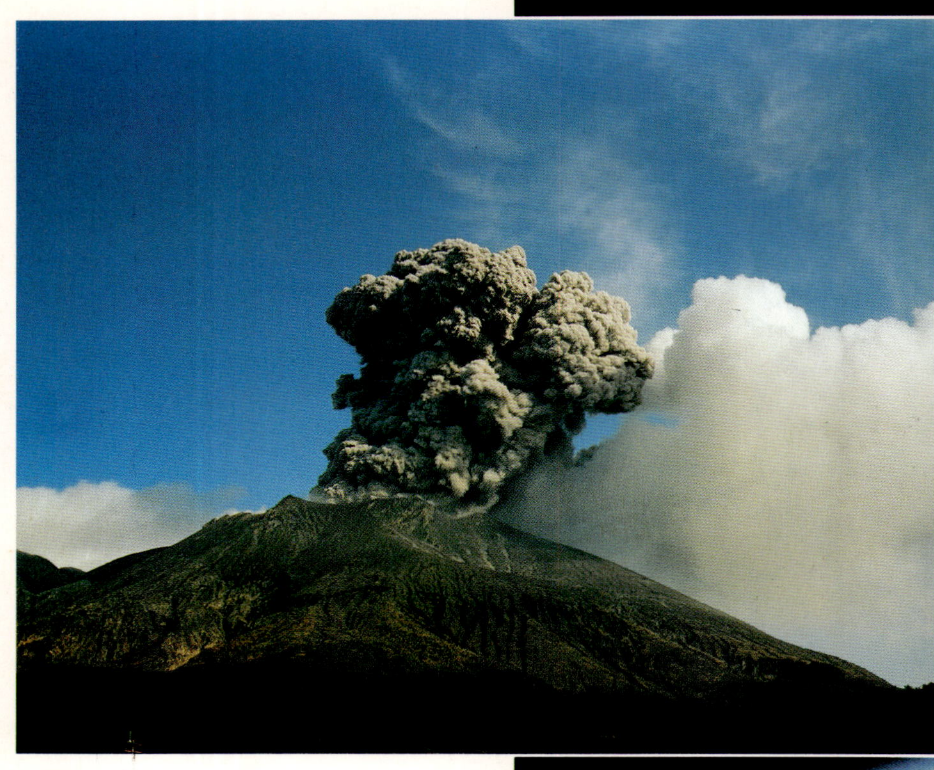

Above: Sakurajima has been almost continuously active since 1955 and has erupted thousands of times. It ejects rocks large enough to break car windows, and at such times, schoolchildren nearby are required to wear helmets outdoors. In November, 1986, a rock weighing five tons was launched more than 2 kilometers into the air. It landed on a hotel roof and dropped right through to the basement.

Right: Japan is located in the zone of high seismic activity around the Pacific rim and is vulnerable to earthquakes and volcanic eruptions. The crustal seafloor plate, sinking beneath the plate that carries the landmasses, is partially melted by the heat of the earth's interior, and molten rock rises to the surface to erupt as volcanoes. In this image, which shows the southern region of Japan, a wind-borne plume of ash can be seen trailing from the volcano Sakurajima (bottom right).

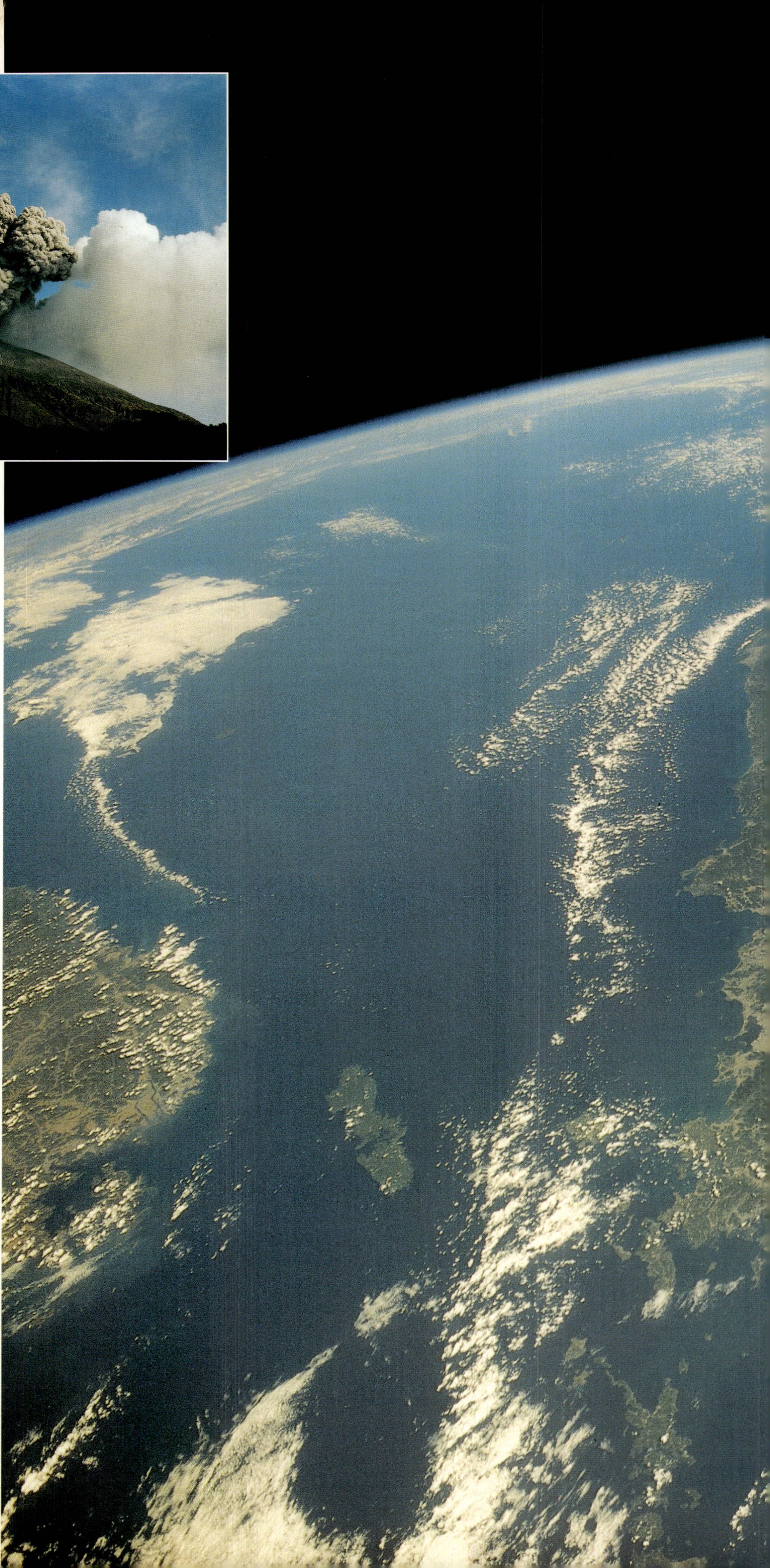

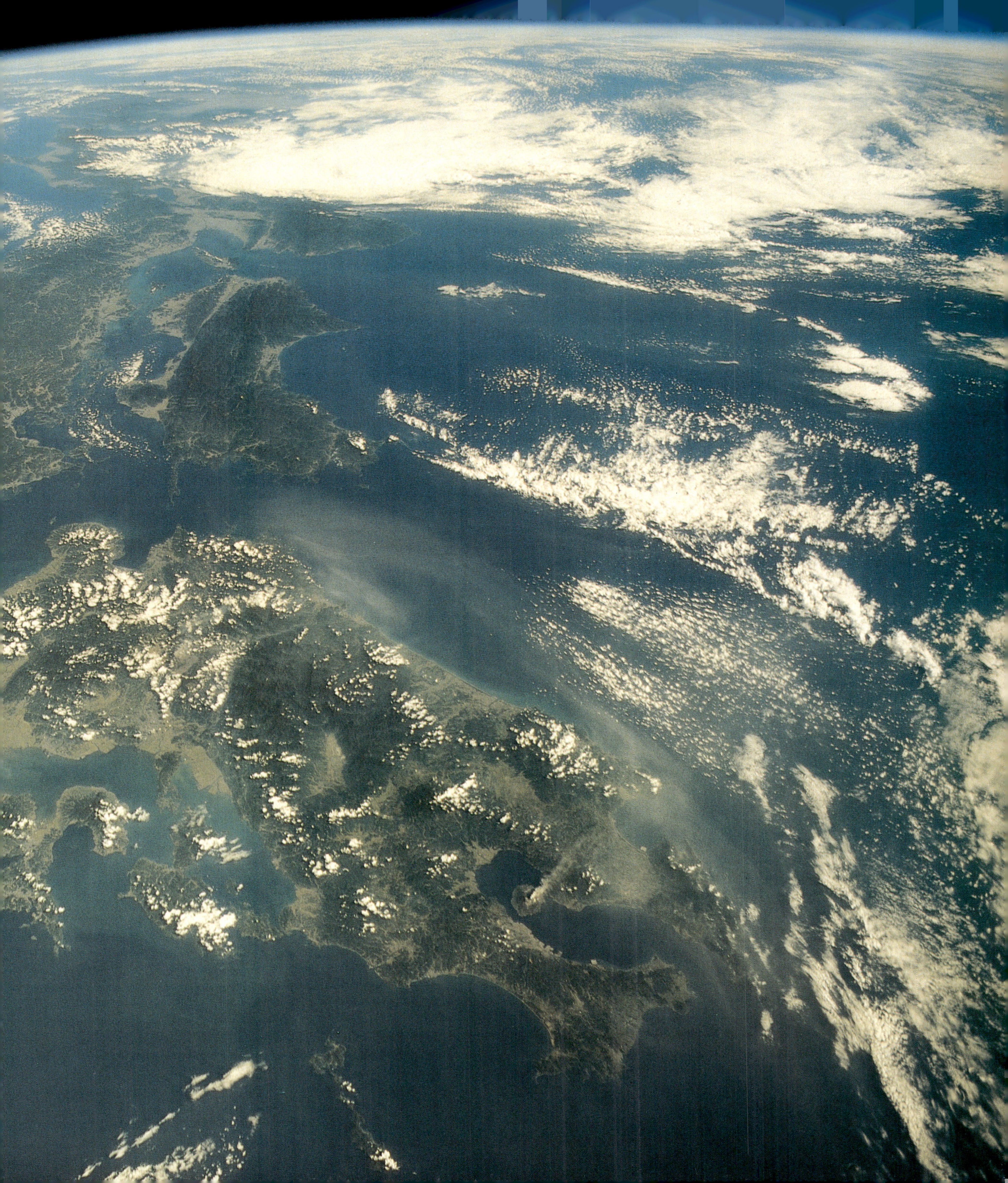

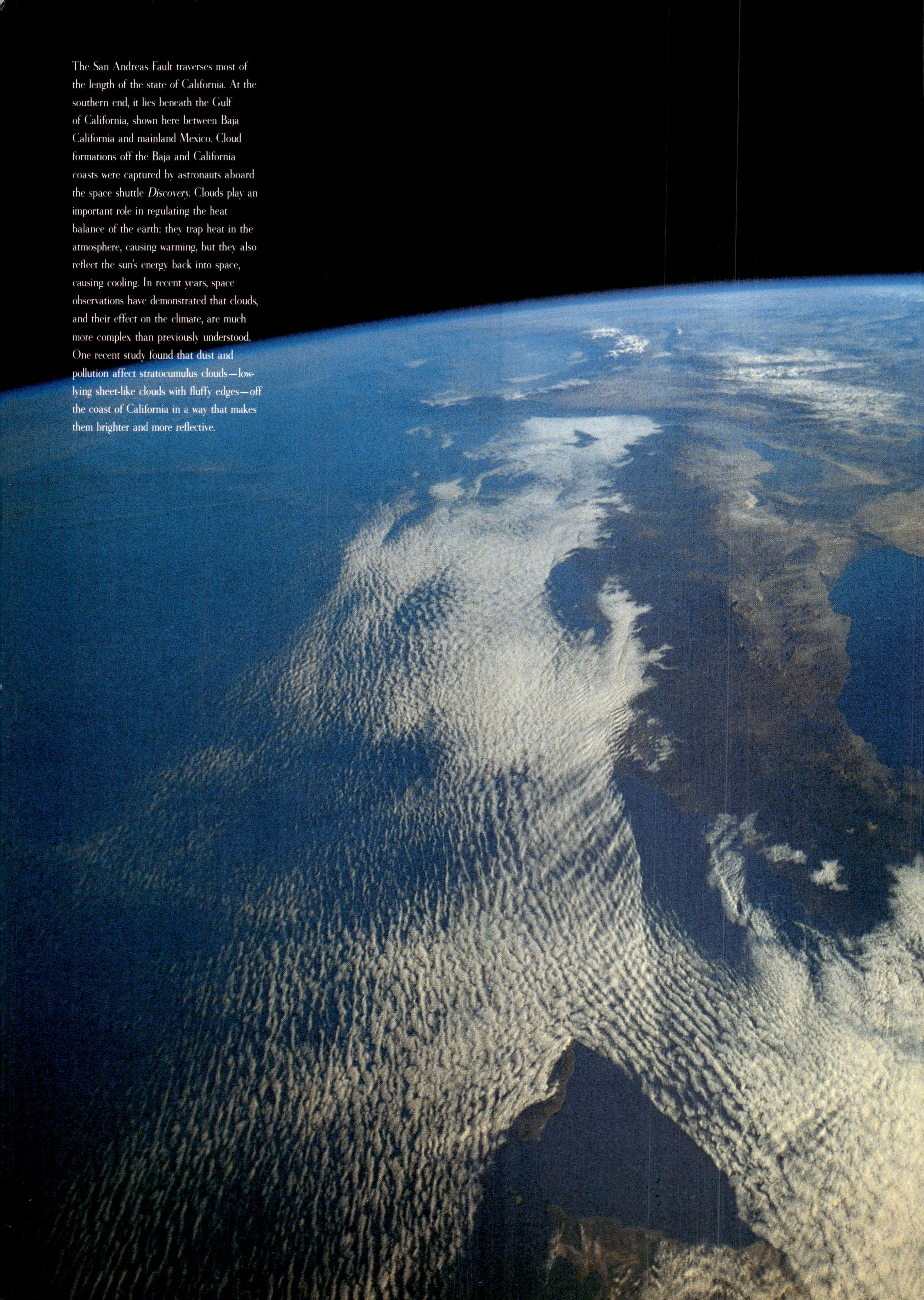

The San Andreas Fault traverses most of the length of the state of California. At the southern end, it lies beneath the Gulf of California, shown here between Baja California and mainland Mexico. Cloud formations off the Baja and California coasts were captured by astronauts aboard the space shuttle *Discovery*. Clouds play an important role in regulating the heat balance of the earth: they trap heat in the atmosphere, causing warming, but they also reflect the sun's energy back into space, causing cooling. In recent years, space observations have demonstrated that clouds, and their effect on the climate, are much more complex than previously understood. One recent study found that dust and pollution affect stratocumulus clouds—low-lying sheet-like clouds with fluffy edges—off the coast of California in a way that makes them brighter and more reflective.

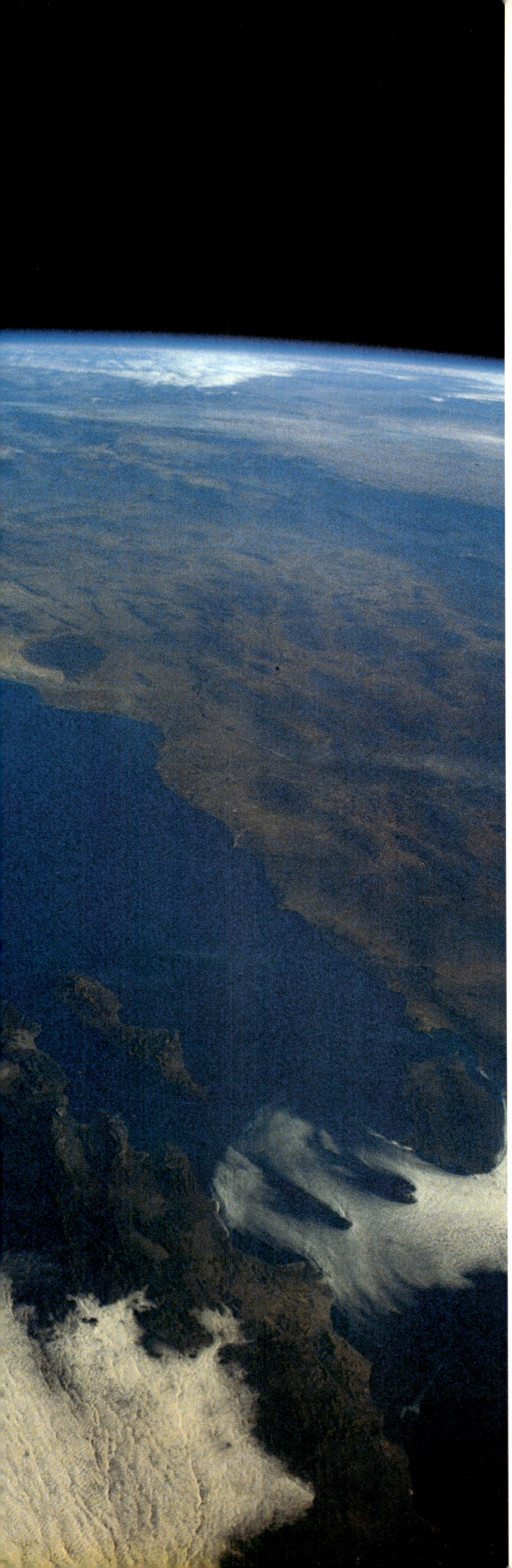

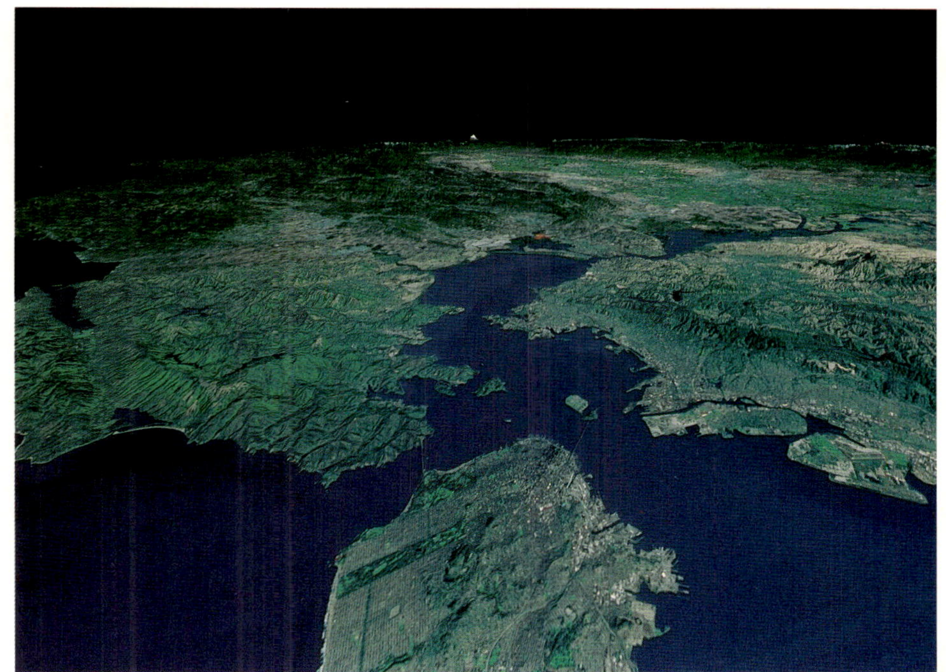

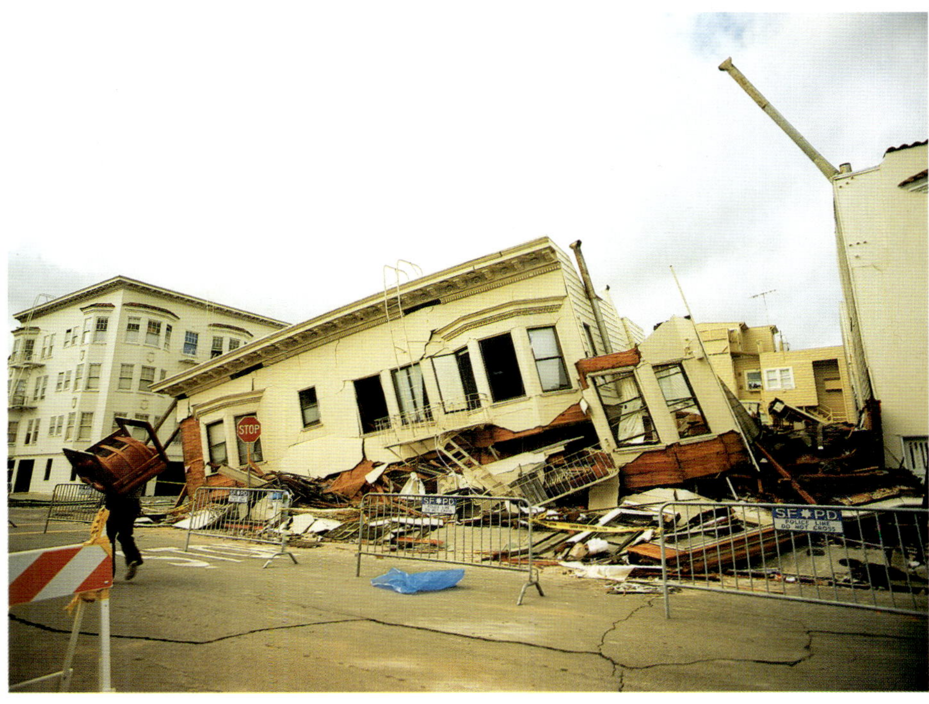

Top: This computer-generated image shows the city of San Francisco (bottom center) flanked by the Pacific Ocean to the left and San Francisco Bay to the right. In *Blue Planet*, this is part of an animated sequence simulating an exhilarating ground-hugging flight along the San Andreas Fault. The simulation, created at the Jet Propulsion Laboratory, required a staggering amount of information—the equivalent of about 9.3 million pages of printed text. It took two large computers eight months of elapsed running time to do the computation for the simulation. Transferring the computed frames to IMAX film almost doubled the amount of data being processed, and it took 259 hours to record the 105-second animation sequence.

Above: San Francisco is located near the San Andreas Fault, the line along which two crustal plates are sliding horizontally past each other. The Pacific plate on the west is moving north relative to the continental plate on the east. The plates slide smoothly at some points, but lock together at others, causing a build-up of strain in the rocks that is eventually released in an earthquake. At 5:04 P.M. Pacific time on October 17, 1989, a quake measuring 6.9 on the Richter scale hit the San Francisco area. The epicenter in the mountains around Santa Cruz was located in a "seismic gap" that had not ruptured in recorded history; two years previously, scientists had suggested it was at high risk for a major quake. More than sixty people were killed and thousands were injured in the quake, which also caused extensive property damage. The devastation was worst in San Francisco's Marina area, which was built on marine muds that amplified the seismic waves four to five times.

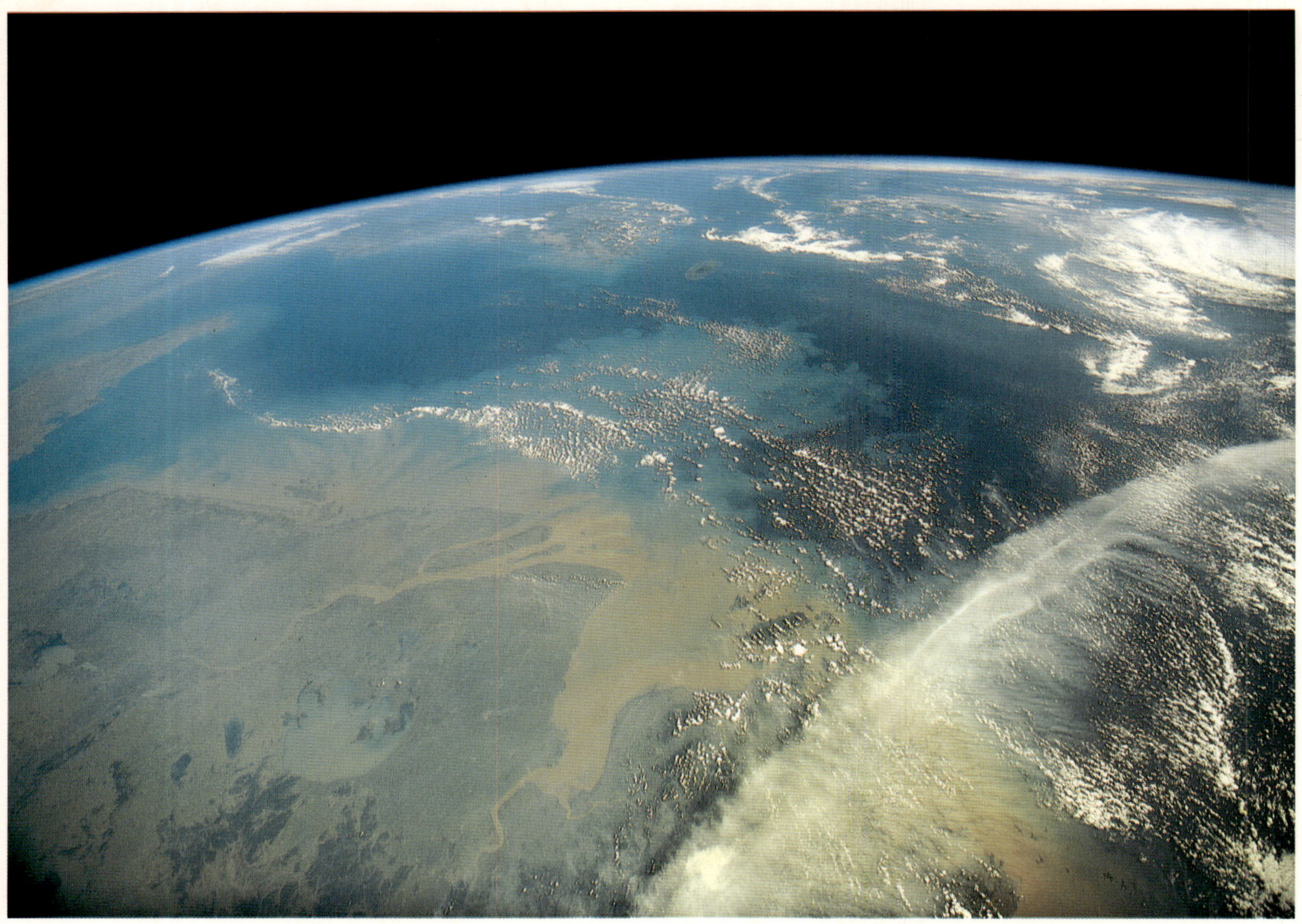

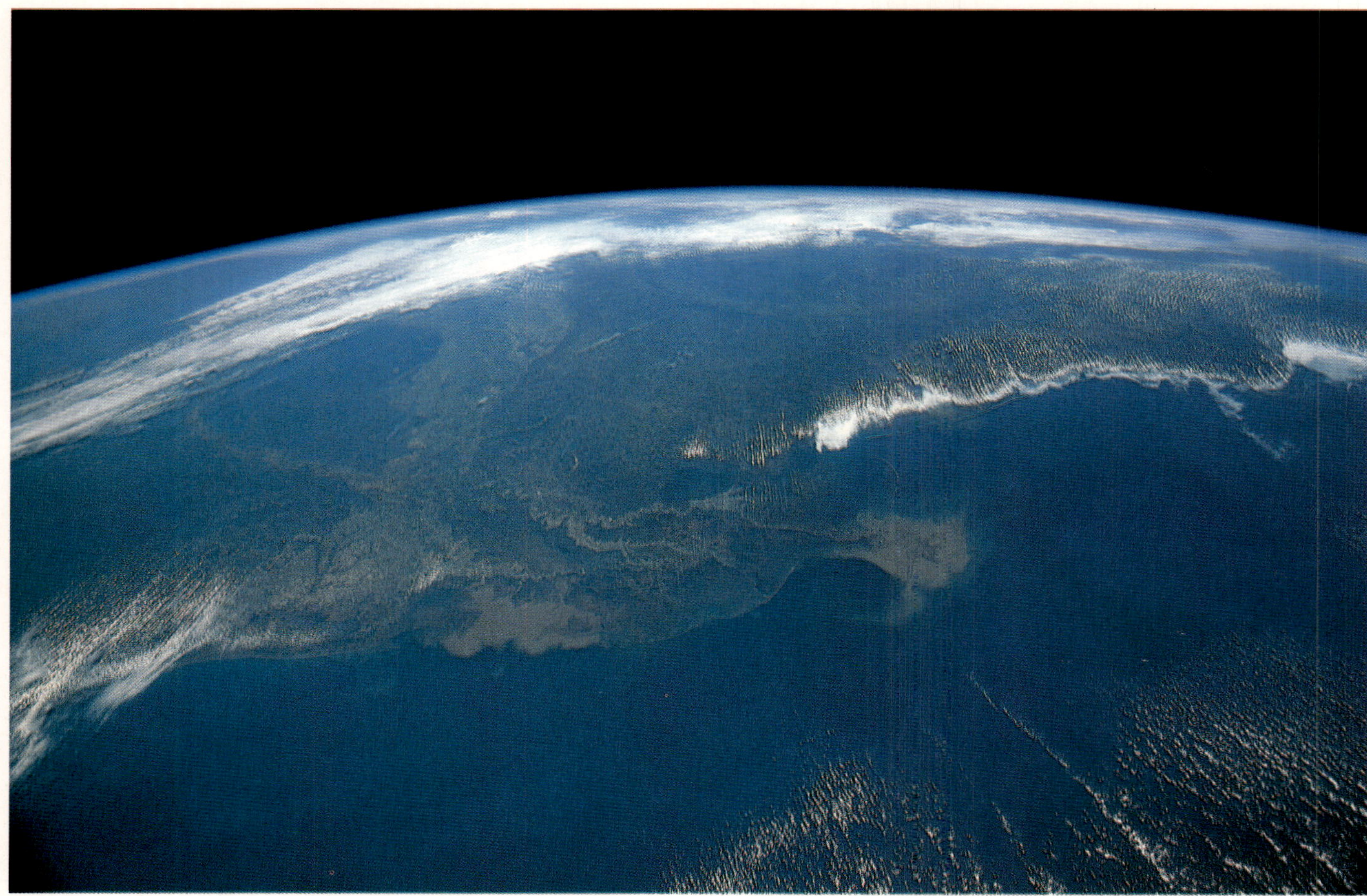

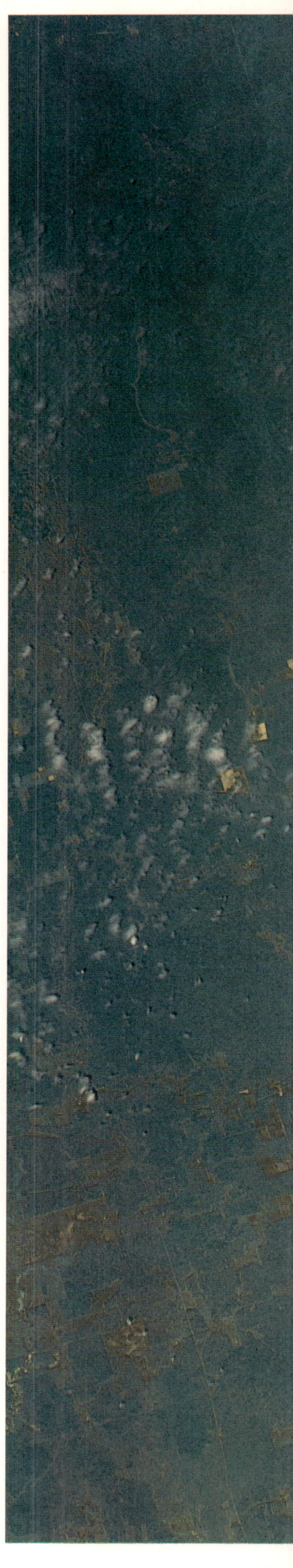

From space, we can see rivers turned brown by silt disgorging tons of valuable topsoil into the oceans. The picture above shows the delta of the Mississippi River, which drains two-thirds of the continental area of the United States, discoloring the waters of the Gulf of Mexico as it carries off soil laced with pesticides. Similarly, the Yangtze River in China (top) ferries a heavy load of silt into the Yellow Sea. It also acts as a dump for industrial wastes and sewage.

Right: The effects of human habitation are clearly visible from space. Here, a patchwork quilt of settlement and agricultural development can be seen on the Chaco Plain, a swampy area covered with small floodplains near the border of Paraguay and Argentina. Trees in this region were cut down many years ago to make way for large plantations growing wheat and other grains.

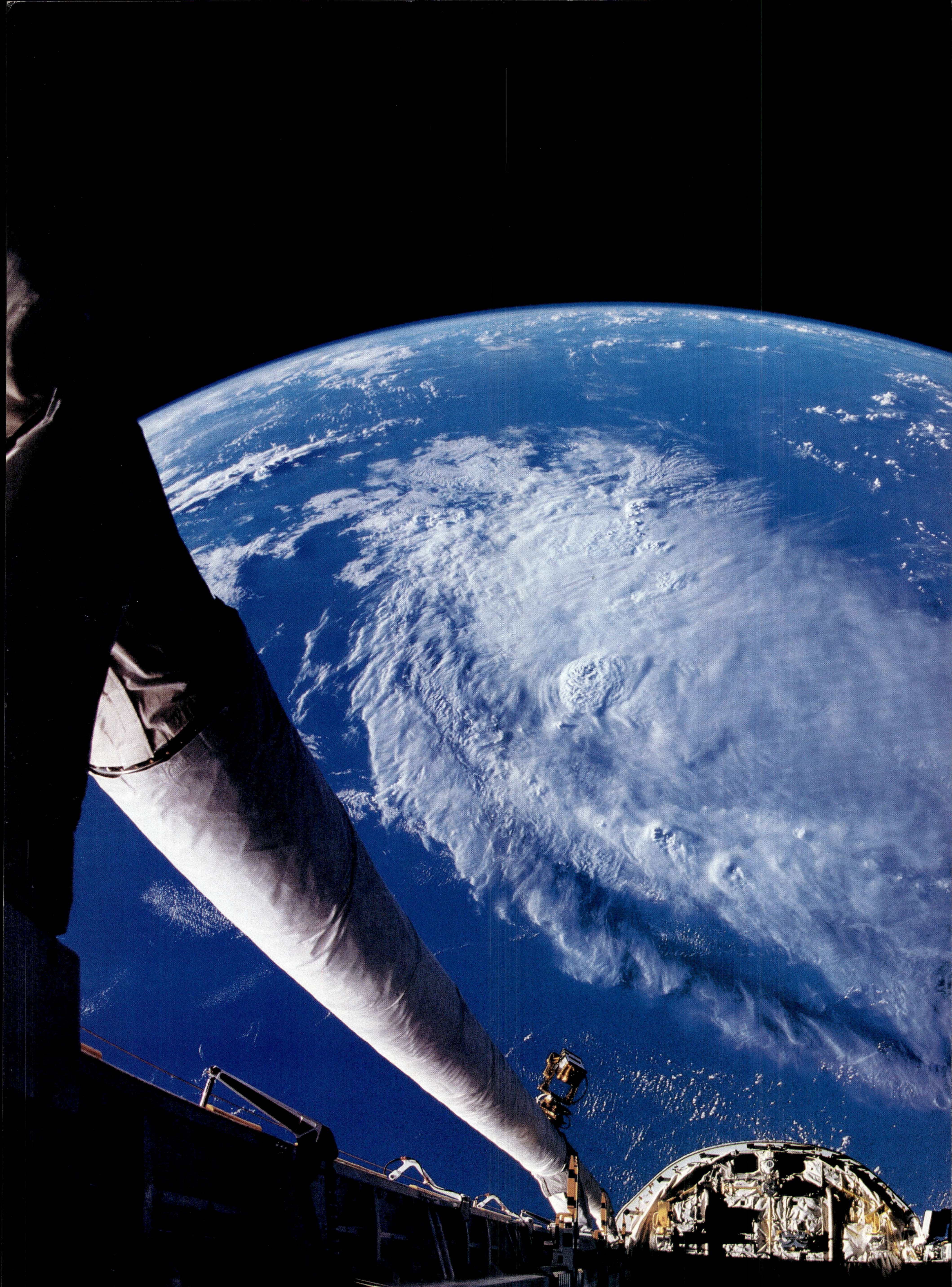

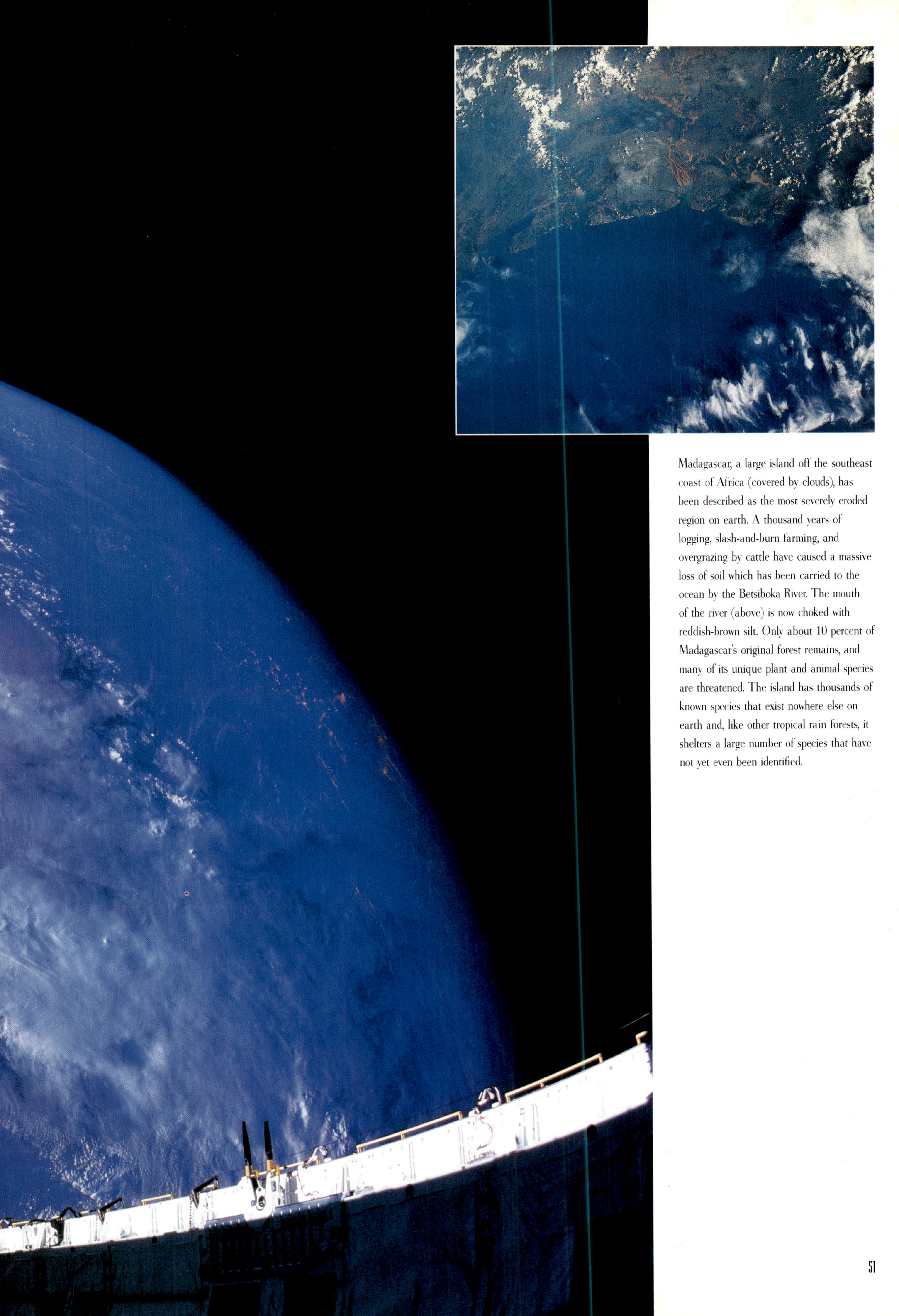

Madagascar, a large island off the southeast coast of Africa (covered by clouds), has been described as the most severely eroded region on earth. A thousand years of logging, slash-and-burn farming, and overgrazing by cattle have caused a massive loss of soil which has been carried to the ocean by the Betsiboka River. The mouth of the river (above) is now choked with reddish-brown silt. Only about 10 percent of Madagascar's original forest remains, and many of its unique plant and animal species are threatened. The island has thousands of known species that exist nowhere else on earth and, like other tropical rain forests, it shelters a large number of species that have not yet even been identified.

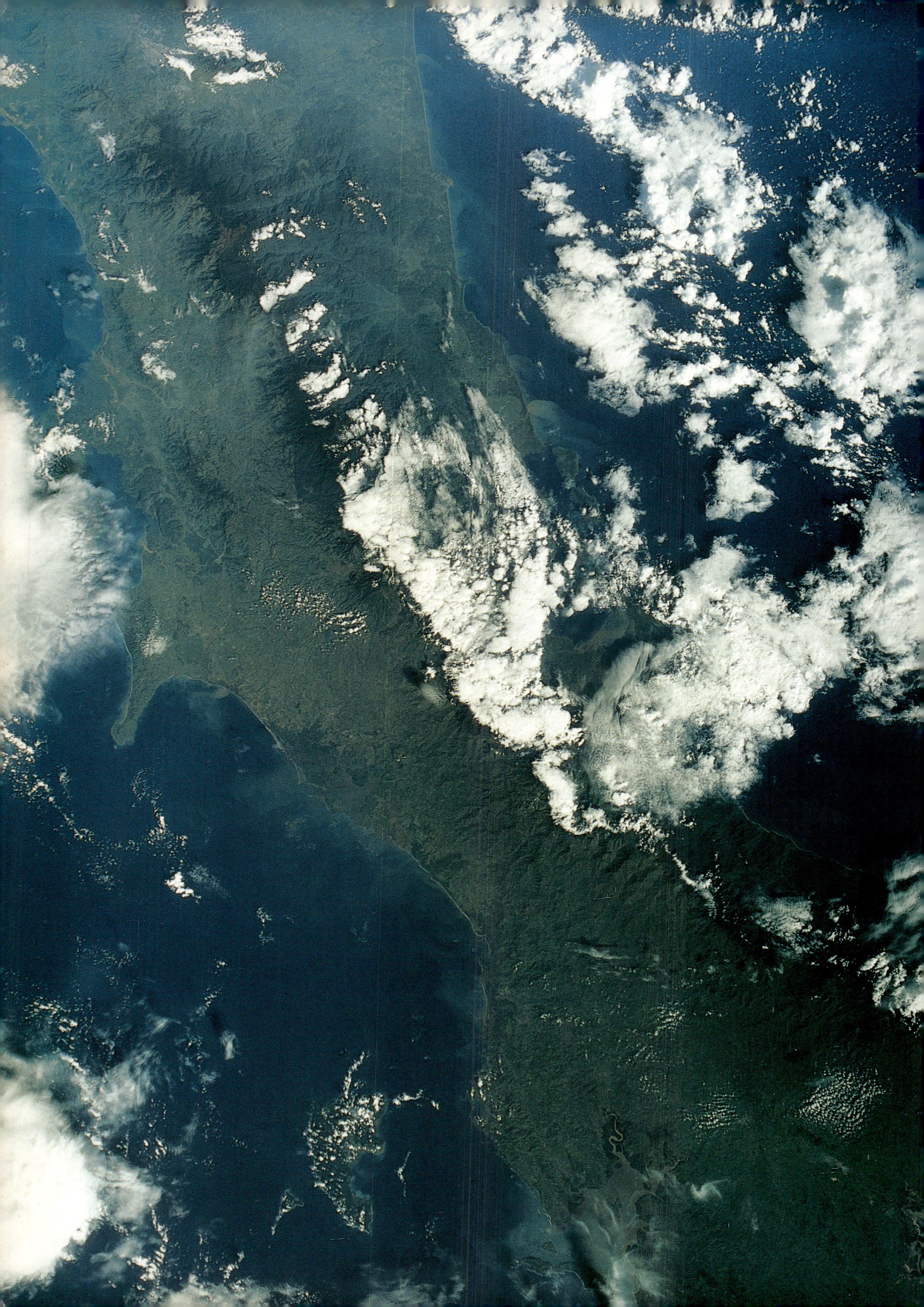

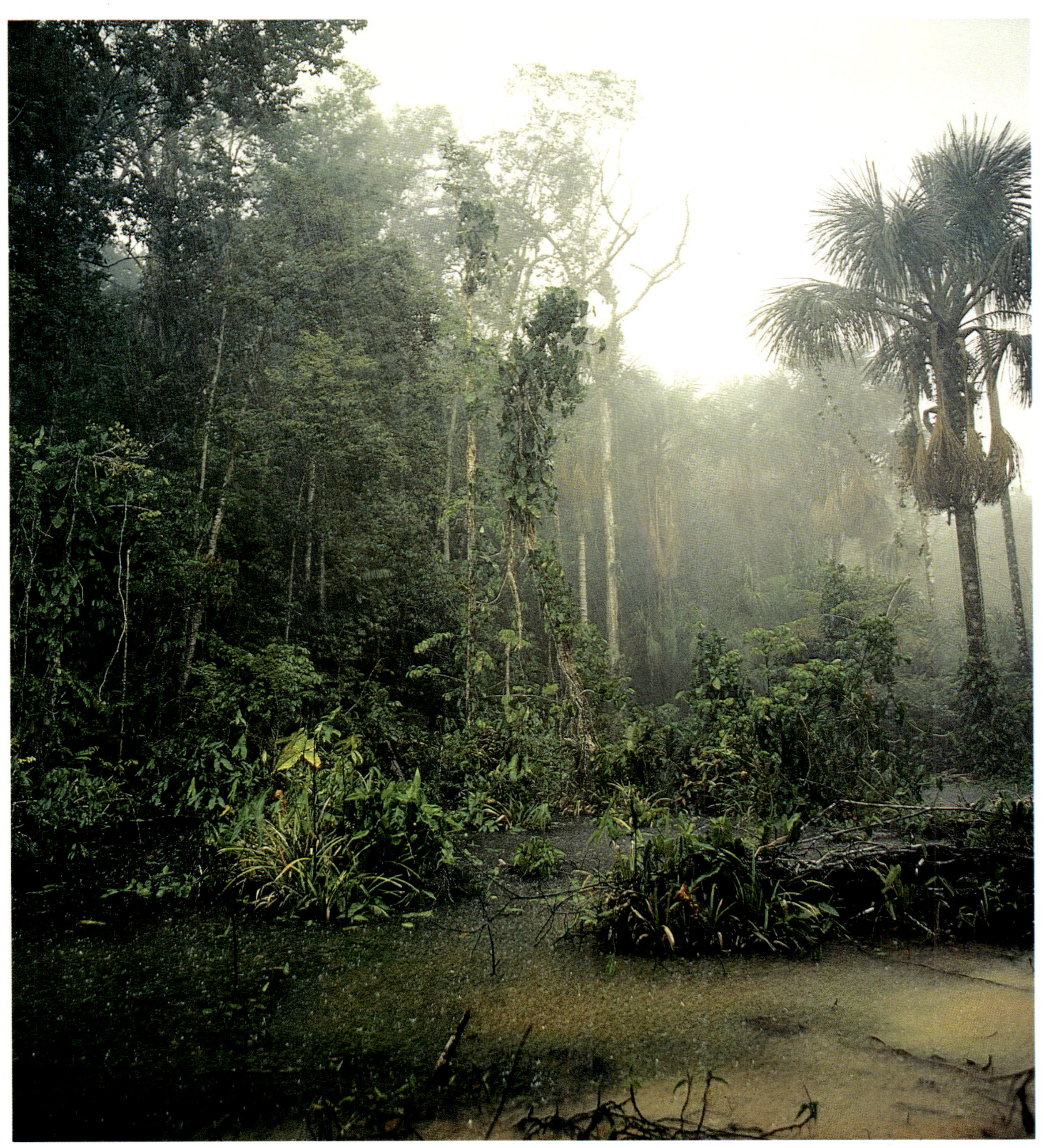

Left: Dense rain forests once nearly covered the Central American country of Costa Rica, but more than two-thirds have been lost. Forest cutting and burning and the growing of cash crops such as bananas, sugarcane, and coffee for export have resulted in erosion, a loss of soil fertility, and contamination from pesticides. As with other tropical areas, the Costa Rican forests provide habitat for thousands of species of mammals, birds, reptiles, amphibians, insects, and plants (including more than a thousand species of orchids).

Above: In the delta of the Amazon River, the second-longest river in the world, there are still vast regions of virgin forest. These trees play a key role in controlling the balance of the earth's atmosphere, taking in carbon dioxide and giving off oxygen. They also lock up carbon in their tissues, making it temporarily inaccessible to the atmosphere. Tropical rain forests are also treasure troves of life-forms, providing habitat for many millions of plant and animal species. Many of these plants are important sources of new medicines.

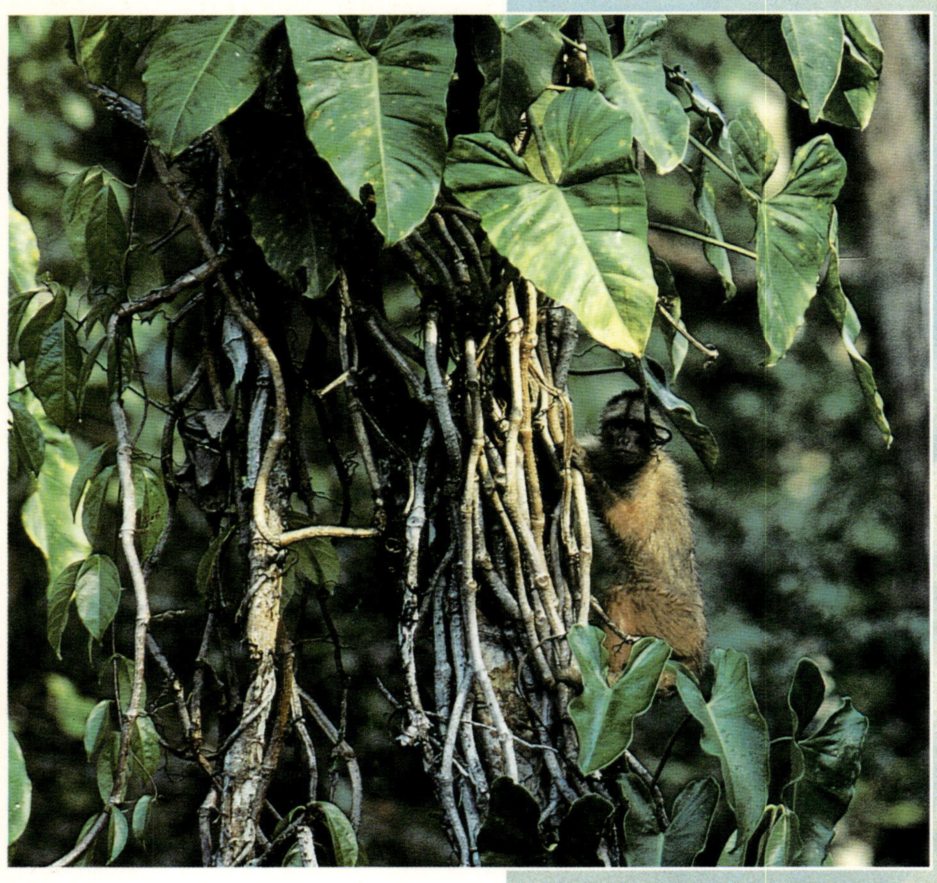

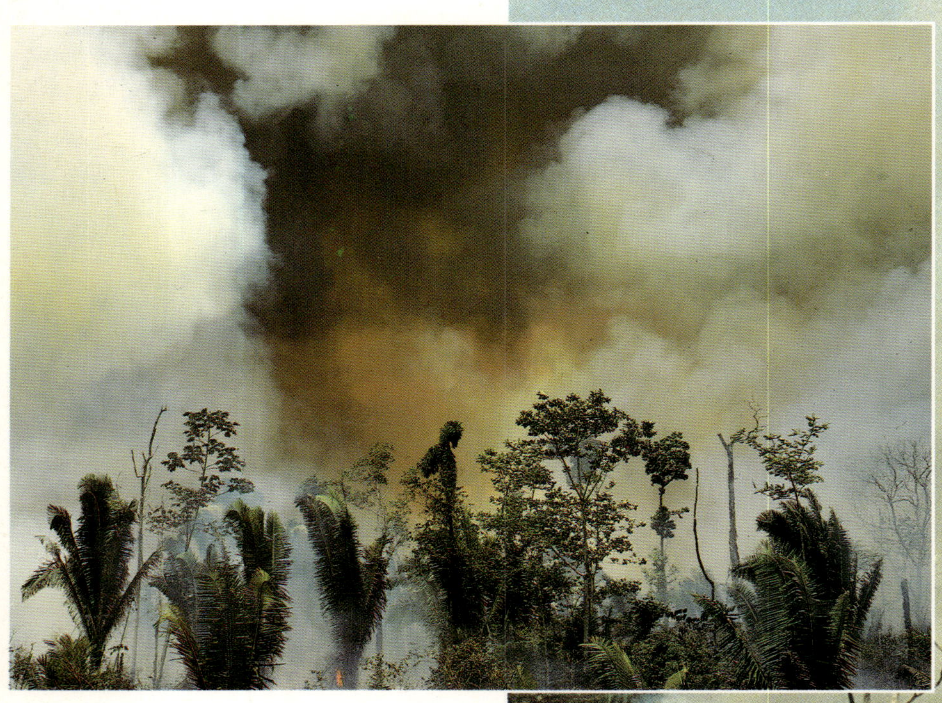

In Brazil, as in other parts of the world, rain forests are being clear-cut and burned to make way for ranching and agriculture. Unfortunately, these operations have not always been successful because tropical soils are often unsuited to intensive farming. Deforestation affects the global climate by releasing large amounts of carbon dioxide into the atmosphere. It also destroys the habitat of millions of species of plants and animals; many become extinct, impoverishing the genetic and biological diversity of life on earth.

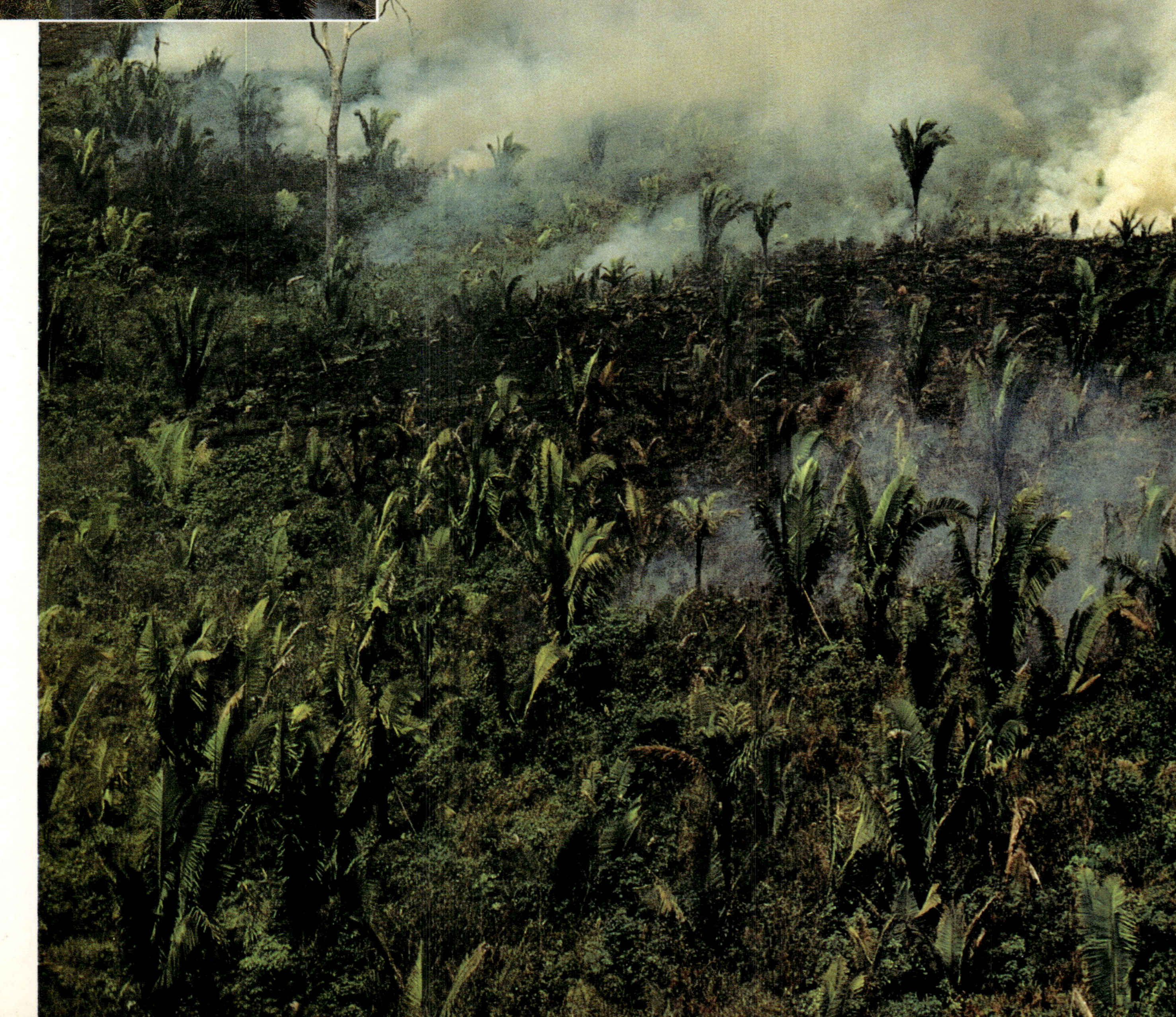

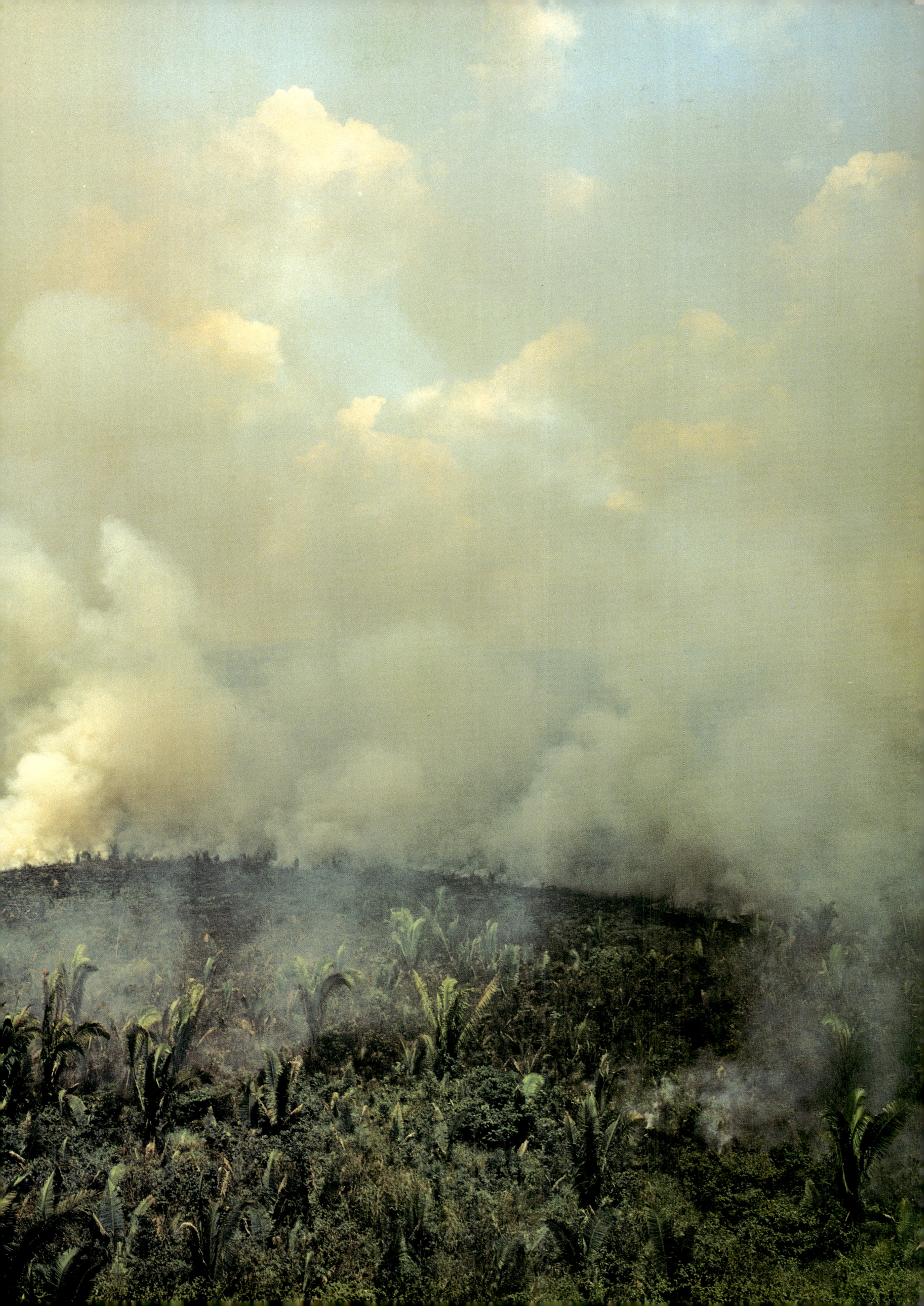

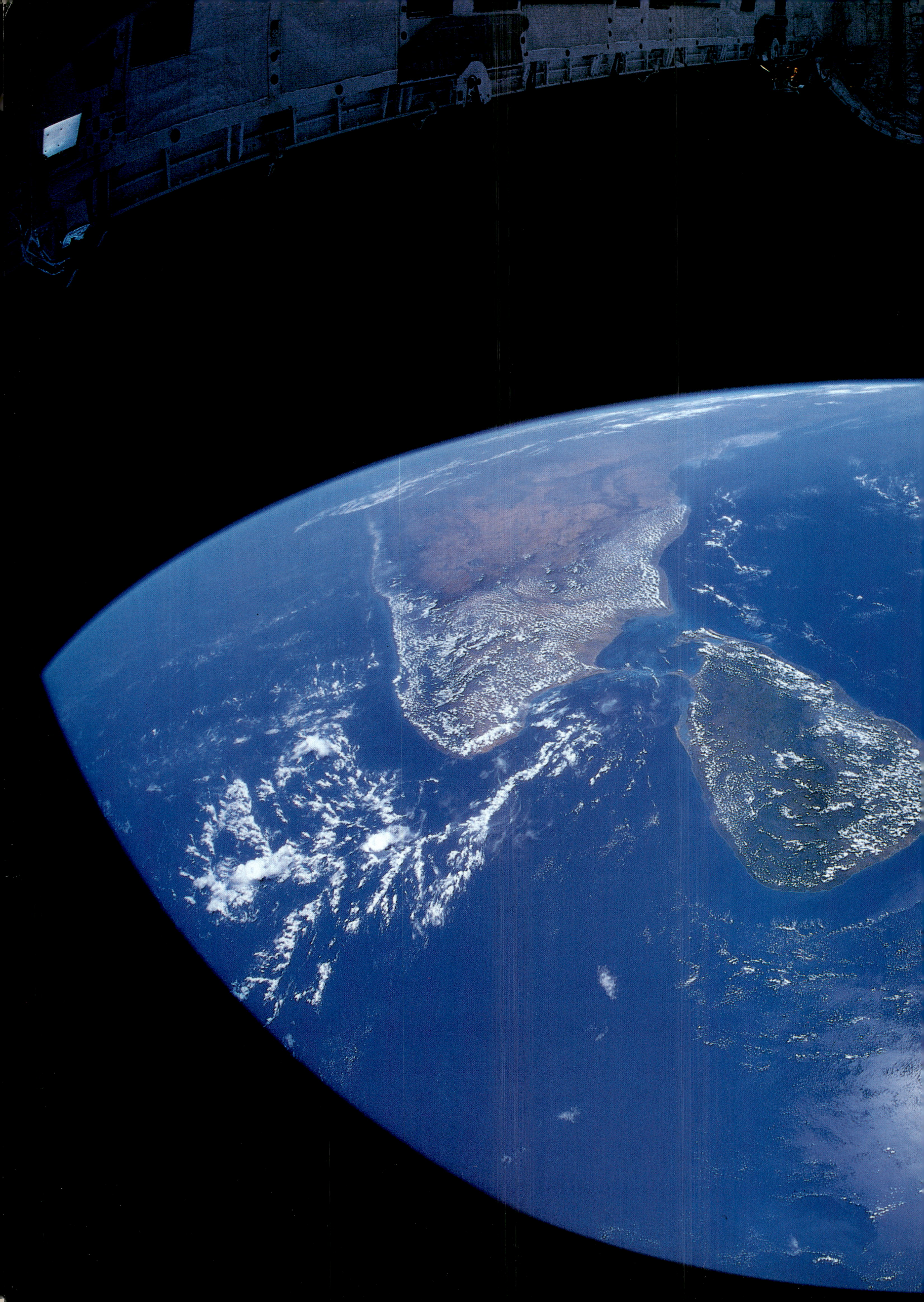

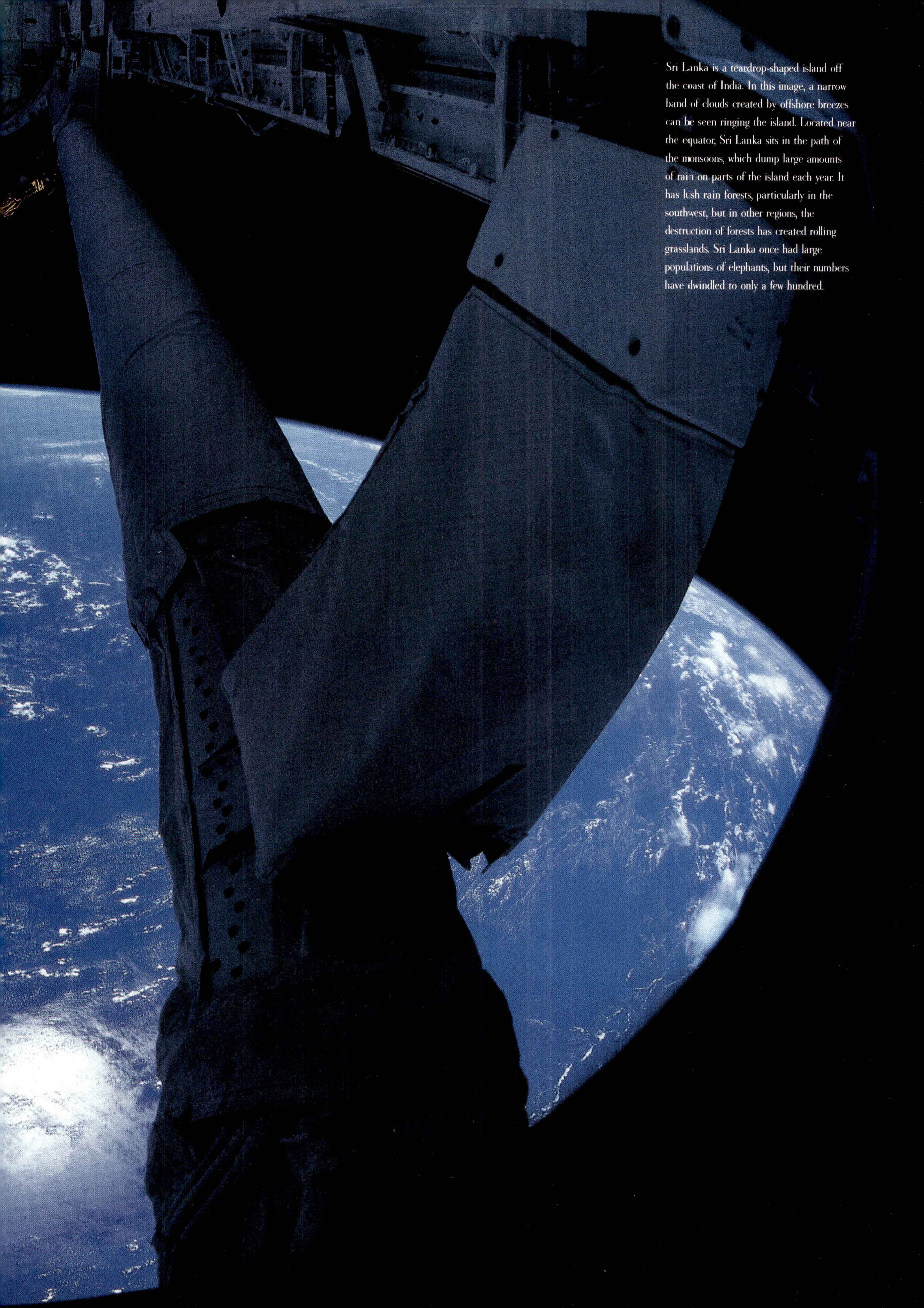

Sri Lanka is a teardrop-shaped island off the coast of India. In this image, a narrow band of clouds created by offshore breezes can be seen ringing the island. Located near the equator, Sri Lanka sits in the path of the monsoons, which dump large amounts of rain on parts of the island each year. It has lush rain forests, particularly in the southwest, but in other regions, the destruction of forests has created rolling grasslands. Sri Lanka once had large populations of elephants, but their numbers have dwindled to only a few hundred.

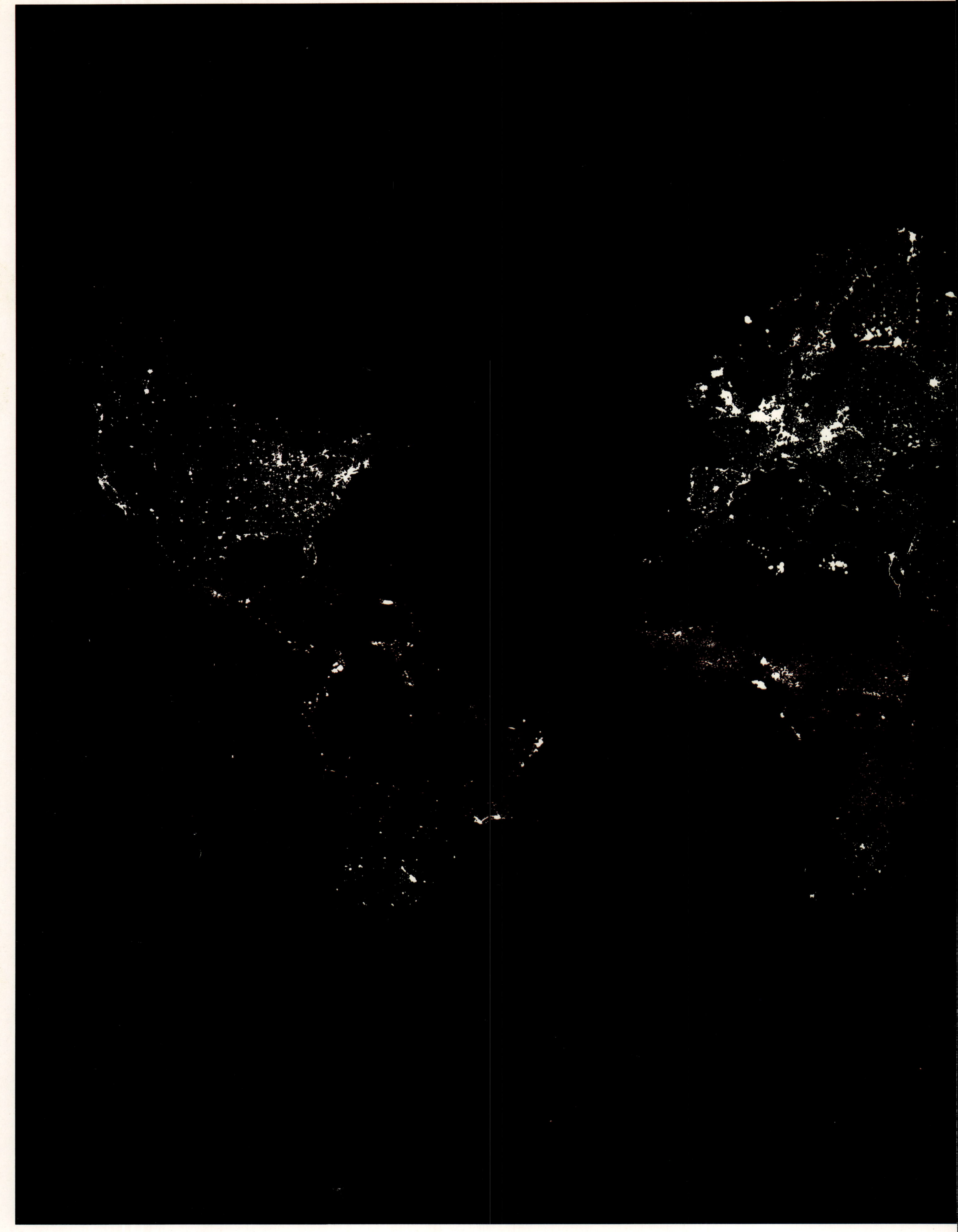

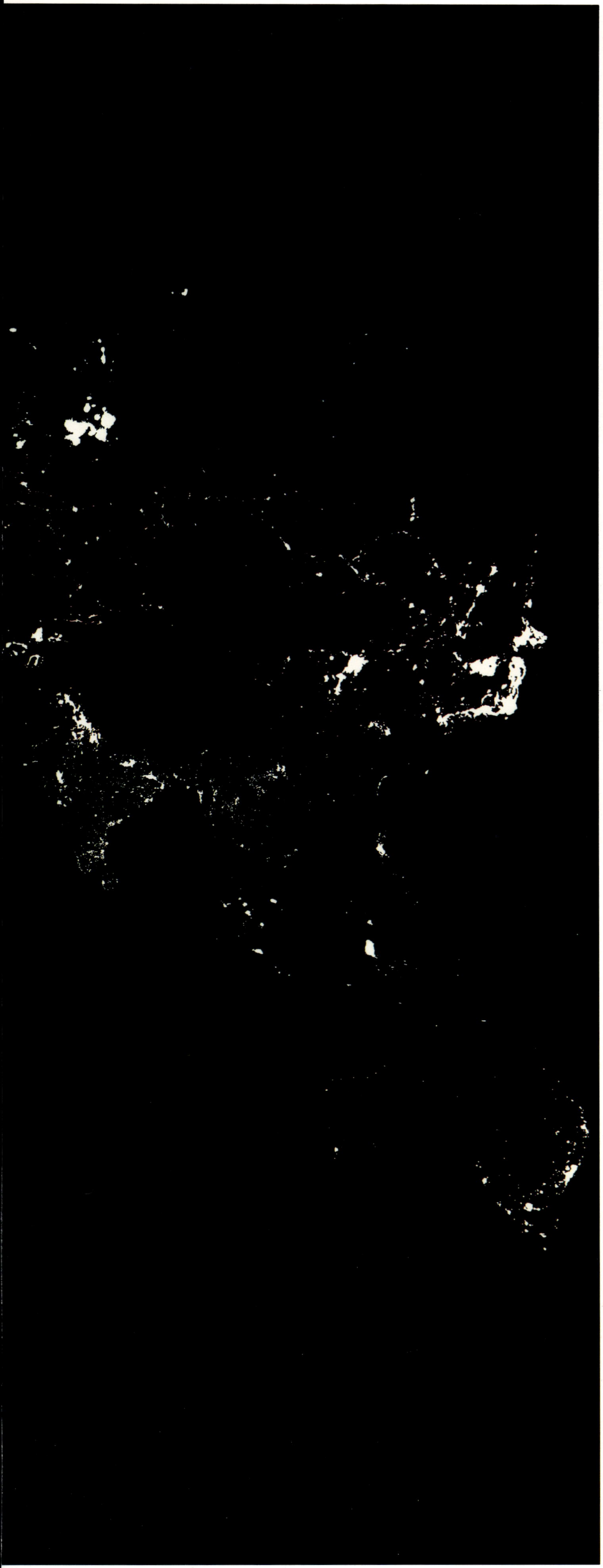

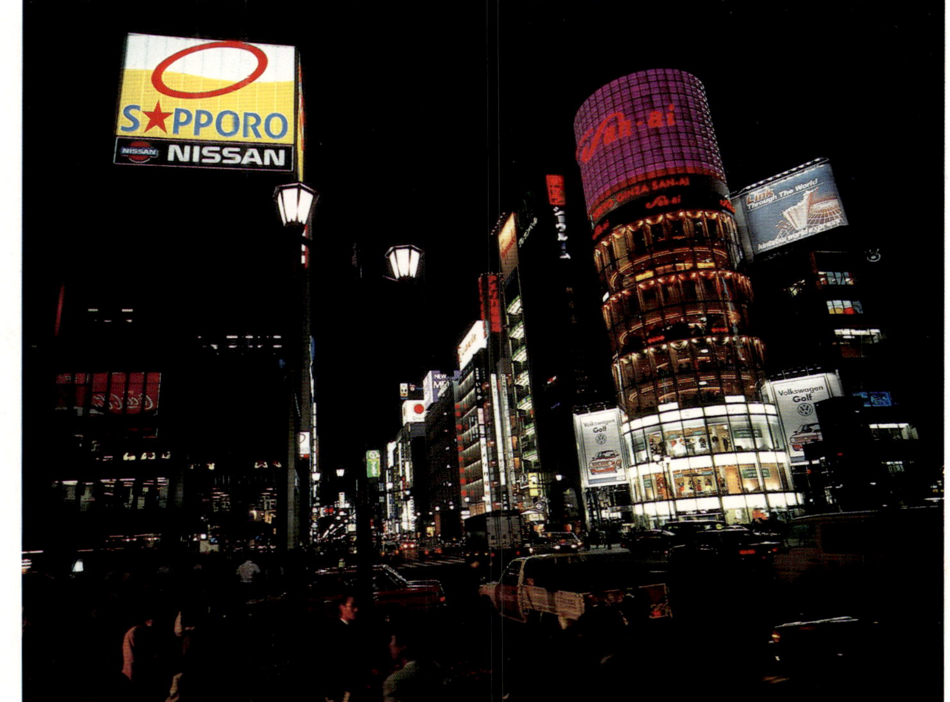

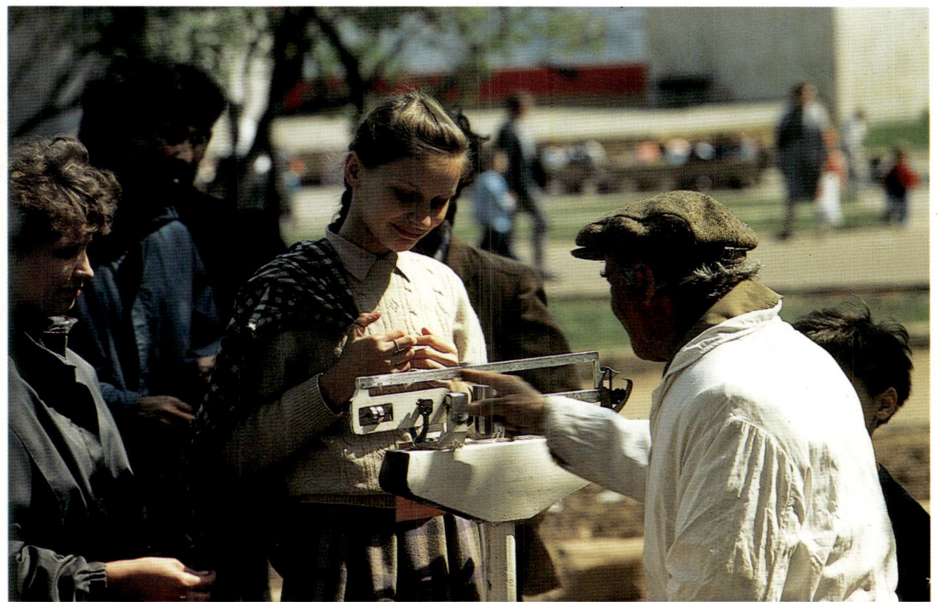

In this light-etched composite image of the globe, we see graphic evidence of the extent to which human activities dominate much of the earth's surface. In the industrialized areas of North America, Europe, the Soviet Union, and Japan, the lights serve as a visual measure of our use of energy. The bright lights in the Sea of Japan are from the Japanese fishing fleet. Gas flares in the Middle East and North Africa are as bright as cities—an indication of how much energy is wasted by flaring. Developing countries are not nearly so prominent; their lights derive more from controlled fires used to burn and clear grasslands and forests for agriculture. Relatively unpopulated areas such as northern Canada, the Australian desert, and the Sahara are dark.

Our most pressing environmental dilemmas are captured in this single image: it documents the profligate use of fossil fuels and the destruction of forests, both of which are contributing to potentially devastating climate warming. This image also catalogues, in a strikingly visual way, disparities between the developed and the developing world in their use of energy and resources.

By the beginning of 1990, there were approximately 5.2 billion people on earth. The global population is now growing by about ninety million people per year. Inevitably, as our numbers increase, so too does the density of the human population—we are increasingly concentrated in large cities. Tokyo, above, and Moscow are prime examples of this trend. In the Soviet Union—the largest country in the world by area—there has been a dramatic shift from a rural to an urban population since World War II. Now, more than two-thirds of the population lives in the cities, which occupy only about 1 percent of the total landmass.

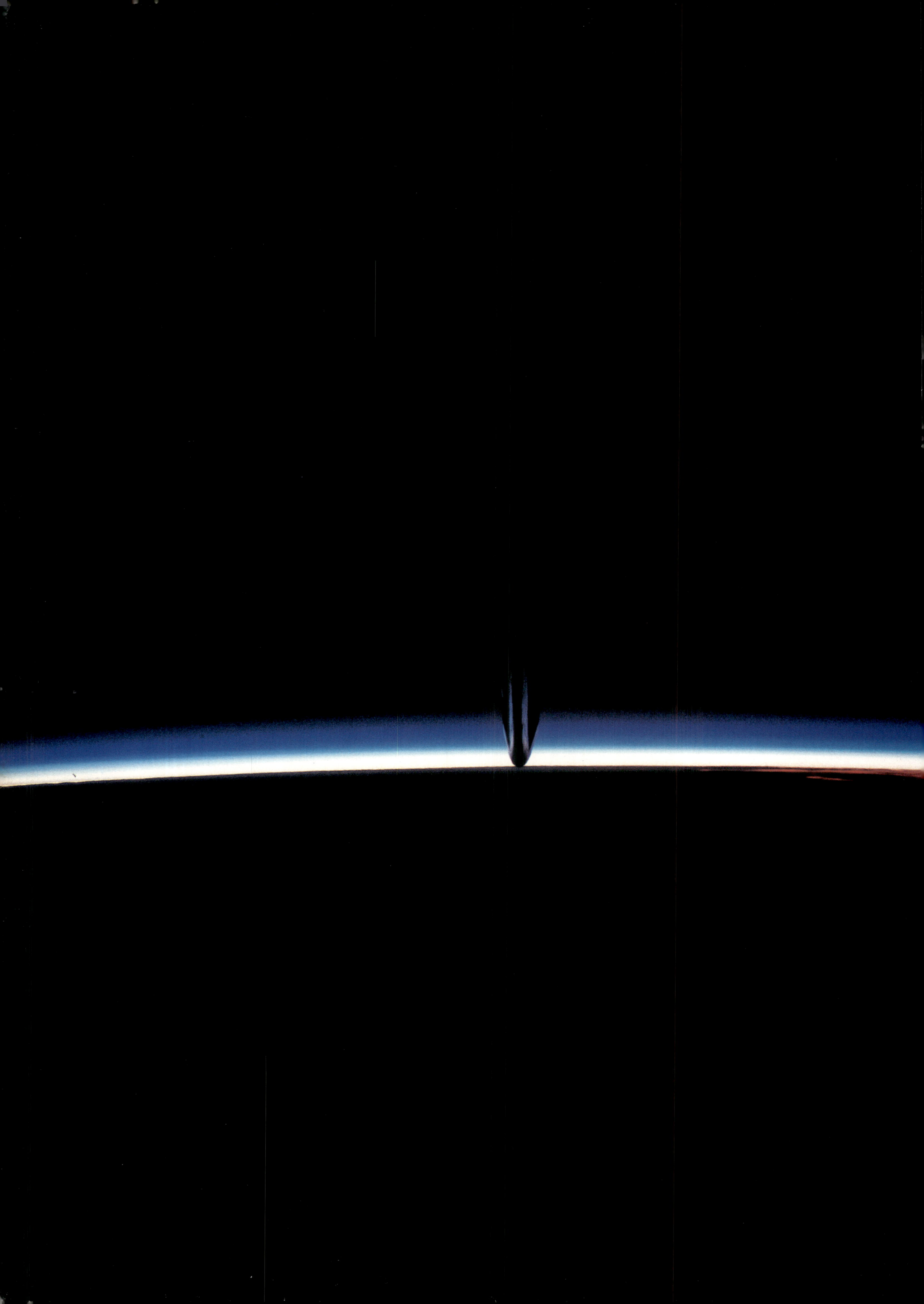

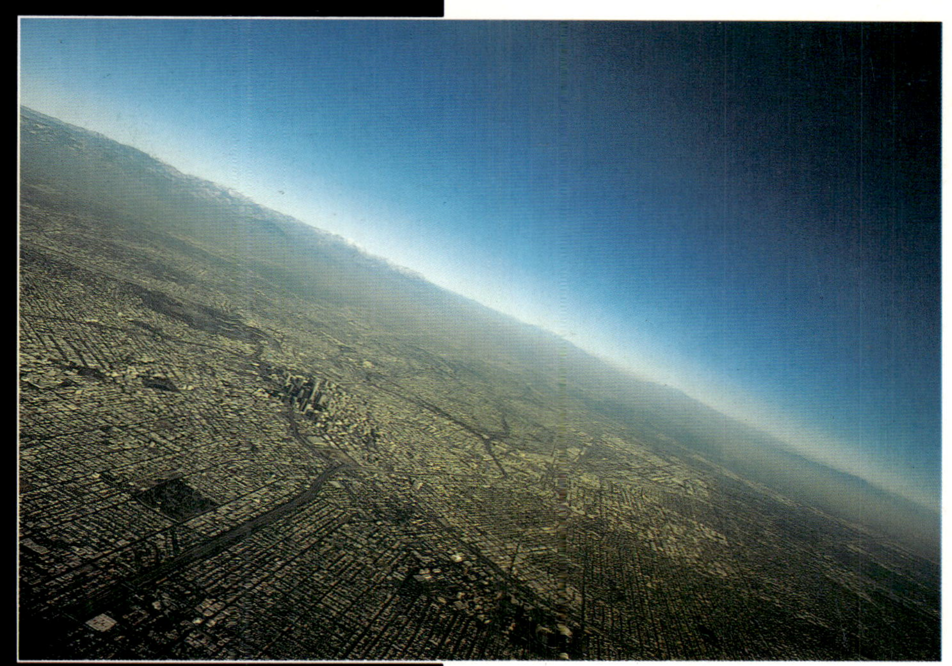

Los Angeles, the second most populous city in the United States, is a city of cars. Sprawling across some 1,200 square kilometers, it is crisscrossed by an intricate network of freeways. Unfortunately, pollutants from car exhausts, when combined with energy from the sun, produce a noxious smog that reduces visibility, damages trees and crops, and causes eye irritations and respiratory problems. Los Angeles is trying to reduce car usage by encouraging carpooling, changes in working hours, and greater use of telecommunications technologies, and by moving offices closer to where people live.

From the vantage point of space, we see that the earth's atmosphere, seemingly so vast, is in reality a thin hazy-blue line standing between us and the deadliness of space. One of its most important components, invisible to the human eye, is ozone, a form of oxygen containing three atoms. Ozone is dispersed tenuously throughout the earth's upper atmosphere between about 15 and 36 kilometers; if it were to be compressed to ground-level atmospheric pressure, it would form a band just 3 millimeters thick. This fragile barrier is all that stands between life on earth and the sun's biologically damaging ultraviolet (UV-B) radiation. If surface levels of this radiation increase, there would be an increase in skin cancer in humans and damaging effects on crops and marine life. The ozone layer has already been reduced by an average of about 3 percent over the whole globe, and large ozone holes now appear over the Antarctic and Arctic at certain times of the year. Ozone depletion is largely caused by industrial chemicals known as chlorofluorocarbons, widely used as refrigerants and foam-blowing agents, and governments are now moving to eliminate their use.

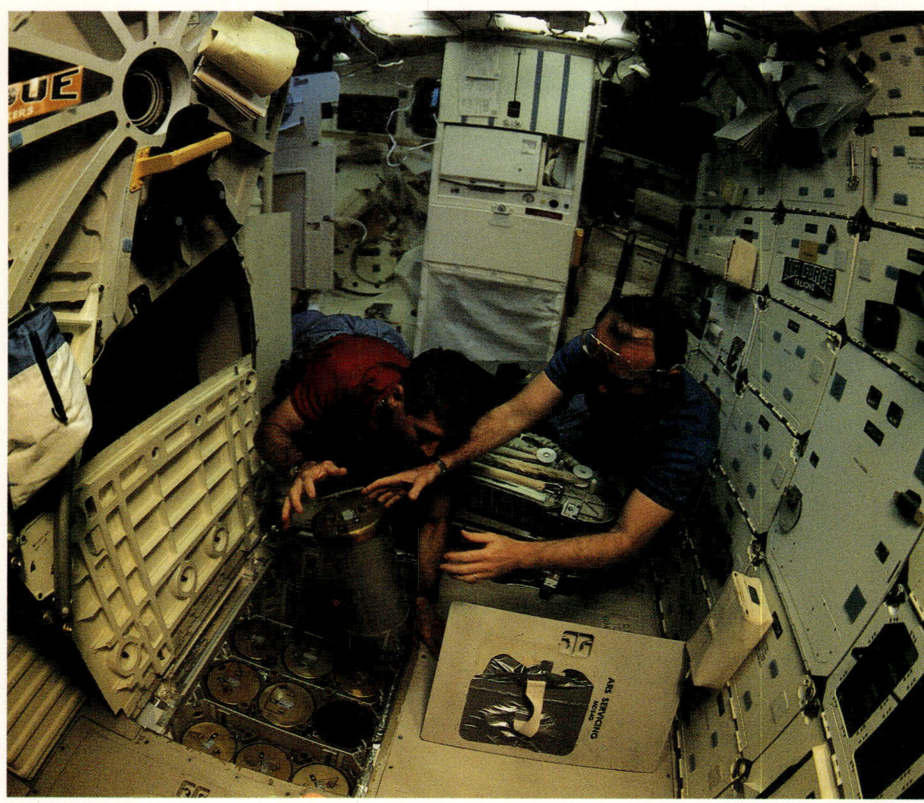

Space provides none of the basic raw materials we need to survive—there is no air, no water, no food—so when we venture out there, we must take our life-support system with us. On a small scale and for limited periods of time, the spacecraft and the space suit provide astronauts with the necessities of survival—just as the earth provides these necessities for all of us. Above, shuttle astronauts can be seen exchanging lithium hydroxide canisters that scrub the spacecraft's atmosphere of excess carbon dioxide. We're extraordinarily careful to conserve and recycle resources on a spacecraft, and to prevent contamination of its life-support system, because failure to do so would have disastrous consequences. Can we claim to be taking equal care with the life-support system that is earth?

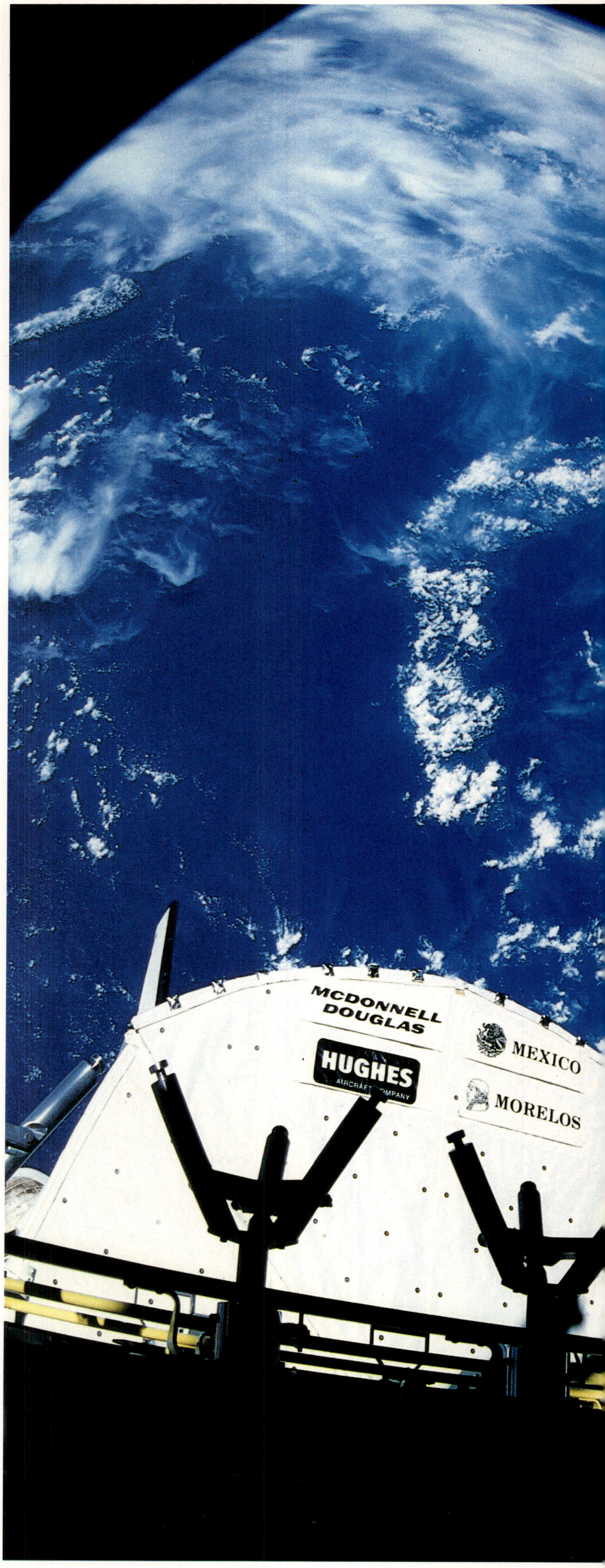

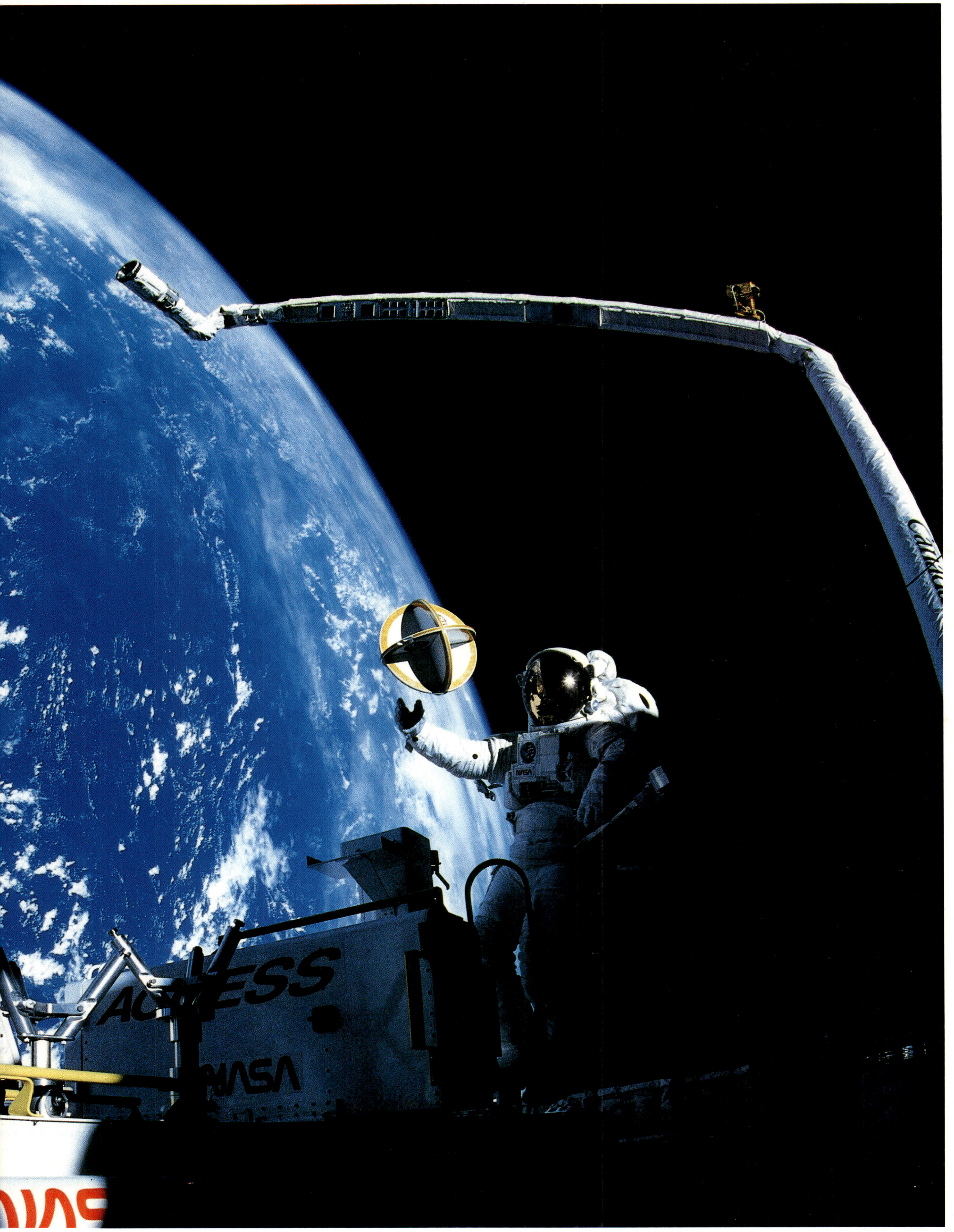

ACCESS
NASA
NASA
Canada

Brilliant hues of blue, green, and turquoise in the oceans give earth its "blue planet" look from space. Here, in the northern Caribbean, we see Jamaica (center left) and, above it, the southernmost part of Cuba. To the right is the island of Hispaniola, shared by Haiti and the Dominican Republic.

Images such as this help us to see our home in new ways—scientifically, emotionally, and philosophically. The fragility of the blue planet has never been more apparent than when it is seen suspended against the vast black backdrop of space.

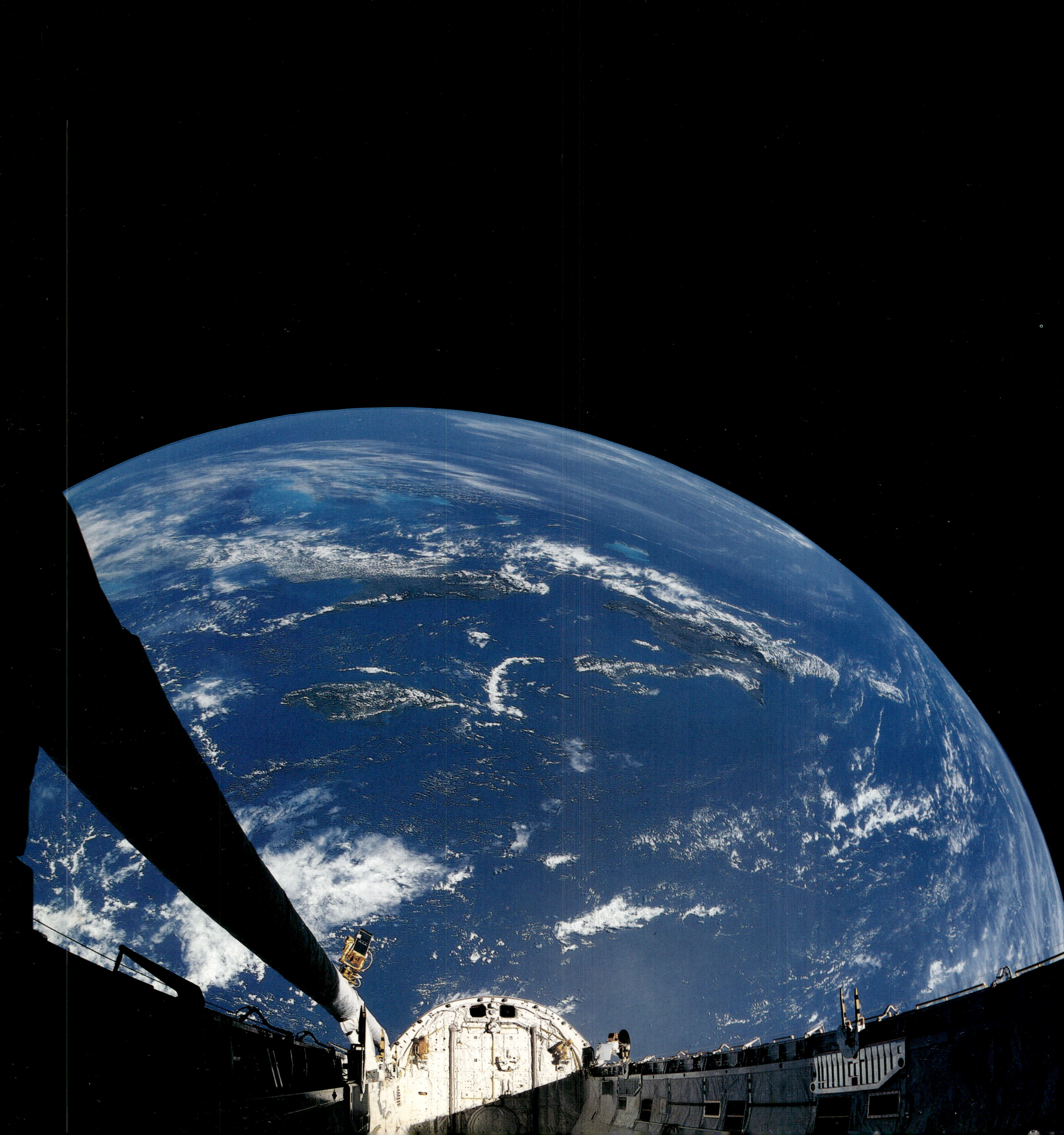